P9-DDO-994

MYSTERIOUS PLACES
THE MEDITERRANEAN

MYSTERIOUS PLACES
THE
MEDITERRANEAN

Philip Wilkinson & Jacqueline Dineen

Illustrations by Robert Ingpen

CHELSEA HOUSE PUBLISHERS
New York • Philadelphia

First published in the United States by
Chelsea House Publishers, 1994

First Printing
1 3 5 7 9 8 6 4 2

Simplified text and captions by **Jacqueline Dineen**
based on the *Encyclopedia of Mysterious Places*
by Robert Ingpen and Philip Wilkinson.

Editor	Diana Briscoe
Designer	Design 23
Art Director	Dave Allen
Editorial Director	Pippa Rubinstein

ISBN 0-7910-2751-1

Printed in Italy

CONTENTS

Introduction

The Mediterranean is one of the most popular places to go for a holiday. One reason for this is its warm, sunny climate; another reason is its history. There are literally thousands of historical sites scattered all around the region and the Mediterranean was the home of the civilisations which have had the most lasting influence on our own culture – ancient Greece and Rome.

Earliest beginnings

It is not surprising that a region so rich in history should contain some of the most ancient religious buildings that have survived to this day. These are the mysterious temples at Tarxien and other sites on the island of Malta. No one knows who the people were who first built them more than 3000 years ago. They did not leave any written records and seem to have died out or left the island. The people who followed them on the island did not continue their traditions.

However, the temples at Tarxien show how long-lived the civilisation of the Mediterranean is. The next place in this book, the palace at Knossos on Crete, is also very old, but it stands at the beginning of a tradition that did continue. In some ways the Minoans who built the palace are almost as mysterious as the ancient Maltese of Tarxien. They did leave some written records behind, but we cannot read all their languages. They also left the beginnings of one of the world's most mysterious stories – the maze or labyrinth.

In spite of all these mysteries, the Minoans of Knossos are somehow close to us. They traded with, and were probably conquered by, the Mycenaeans, who were warriors from mainland Greece. These warriors, though very different in outlook, were the ancestors of the Classical Greeks of the fifth century BC.

The classical civilisations

In this book the classical civilisations of ancient Greece and Rome are not represented by their famous centres – Athens and Rome – but by more mysterious places. For Greece, we have selected two centres of ritual and art – Delphi and Epidaurus – to which people flocked from far and wide. Delphi especially was a popular place of pilgrimage. One reason for this was that to the ancient Greeks it was the centre of the world.

This idea shows the importance of the Mediterranean area in the ancient world. Today we might think of the world's centres as being New York, London or Tokyo, but in ancient times it was the cities around the Mediterranean that seemed to be at the centre.

The sea was a great highway of the ancient world. Peoples such as the Phoenicians (who came from the eastern Mediterranean) and the Greeks were great sailors, and their ships sailed all around the Mediterranean on missions of trade and conquest. The Phoenicians established many trading posts along the coast of northern Africa. It was in one of their colonies that the Romans built their city of Leptis Magna. Leptis was the opposite of Delphi – a place to which very few people came; it became a 'lost city' soon after it was built. It shows that a great empire must also have its outposts, and that these can be as magnificently planned and built as the cities at the centre.

The focus moves eastwards

After Rome and its satellite cities like Leptis Magna had had their heyday, the focus of the Mediterranean moved to the east. Here commanding the Bosphorus, the narrow strait that connected the Black Sea to the Mediterranean was built the great city of Constantinople. This was the capital of the empire.

In 1456 it was conquered by the Turks, who eventually gave it its modern name of Istanbul in 1930. The great church of Hagia Sophia and the palace of Topkapi represent these two phases of this city's existence – each magnificent, each with its own distinct way of life and its own mysteries.

Philip Wilkinson

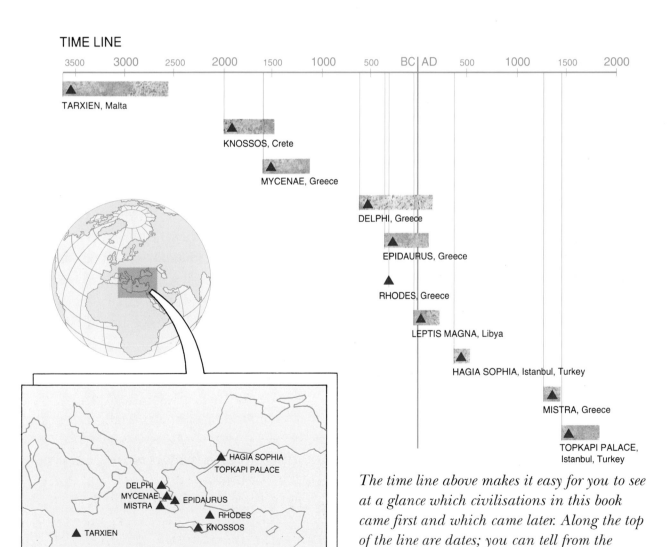

The time line above makes it easy for you to see at a glance which civilisations in this book came first and which came later. Along the top of the line are dates; you can tell from the length of the bars below how long or short a time the site was important. Look at the map to see where the sites are in relation to each other.

Tarxien

Malta, c. 3600–2500 BC

What are the mysterious buildings on Malta?
Are they Europe's first real temples? And what do we know of the Stone Age people
who built them more than 5,000 years ago?

Over 5,000 years ago a group of Stone Age people were living on the island of Malta. They lived an isolated life, even though their island was only 50 miles (80 kilometers) south of Sicily and Italy. They did not learn to write so they did not leave any records; we do not even know what they were called. Yet, in their own way, these long-dead people were unique. Why was this?

The answer lies in the mysterious buildings they left behind. They are unlike any other buildings anywhere else in the world. Archaeologists think that they are temples. If that is so, these people must have developed the first temple-based religion in western Europe. The buildings were used for over 1,000 years and then these unknown people disappeared. So what happened to them?

Everything we know about early settlers on Malta comes from evidence found on the island. The first people to settle there probably came from Sicily in about 5000 BC. They grew crops such as wheat, barley, and lentils, and kept cattle, sheep, and goats.

Monumental feats

By the time these people put together their mysterious buildings around 3600 BC, they seem to have learned quite a bit about architecture. The buildings are scattered all over Malta and the nearby island of Gozo. They have obviously been carefully planned and they are all similar.

If they are temples, what made these people decide to build them? Which gods were worshiped there? And, most incredible of all, how did the people build the temples with simple stone tools? They are made from huge stone blocks – or megaliths. How could they move large stones

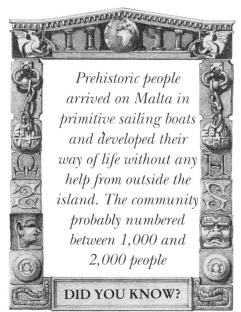

Prehistoric people arrived on Malta in primitive sailing boats and developed their way of life without any help from outside the island. The community probably numbered between 1,000 and 2,000 people

DID YOU KNOW?

Inside one of the temple buildings an islander grinds corn at a communal quern, perhaps as part of a religious ceremony praying for a rich harvest the following year. At the back, people make offerings to an animal statue.

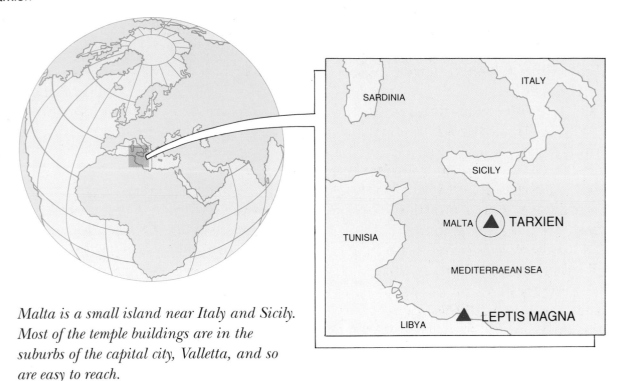

Malta is a small island near Italy and Sicily. Most of the temple buildings are in the suburbs of the capital city, Valletta, and so are easy to reach.

into place without any equipment?

The buildings are in ruins today but archaeologists can work out what they must have looked like. The most spectacular ones are at Tarxien and Hal Saflieni, on the outskirts of Valletta. The building at Tarxien is larger than most of the others. It consists of separate chambers linked together by short, narrow passages.

What were these strange buildings? Archaeologists are fairly sure that they were not houses because they have not found any household items such as pots and pans among the ruins. There are no human bones, so they were not tombs. Archaeologists have found statues which could represent god-figures. And, most importantly, the layout of the buildings follows the design of earlier religious structures.

From tomb to temple

The earliest buildings on Malta are not really buildings at all, but chambers cut

directly into the rock. These chambers are made up of several rooms linked together, like the later temples. Archaeologists have found human bones in the rock chambers, so they must have been tombs. The earliest are very simple and were only big enough to hold a body, but later tombs are bigger and more elaborate.

Archaeologists think that these larger tombs were also used as simple temples where ceremonies for the dead took place. If this is true, the later buildings must represent the next step forward – temples in their own right, where people came to worship.

Who were their gods?

The statues found in the temples are the only evidence of worship. Some of them are similar to "earth mother" figures found in other parts of Europe, which suggests a link with the land. This idea makes sense. The people were farmers, so everything depended on whether their crops would grow. Like many early peoples, their prayers would be for sunshine and rain to give a good harvest.

There is another sign that religious ceremonies centered round farming. At one of the temples at Kordin, archaeologists found a long stone with seven deep bowls cut into it. It looks as if it could have been a quern, which was used for grinding grain.

Why would people grind grain in a temple? Perhaps grinding grain in front of the gods was part of a ceremony to make sure the next harvest was good.

Unknown rituals

Archaeologists have found animal bones in the temples, which probably means that the people made sacrifices to their mysterious gods. Of course, everyone who ate meat left bones lying around, and people sometimes used bones to make tools. But in the temples bits of bone have been hidden away in holes in the walls. Perhaps this was also part of some religious ceremony. It is another of the mysteries surrounding these strange buildings.

There must have been priests to conduct the ceremonies. Priests were always powerful people in the ancient world. People believed that they had contact with the gods and so could control the soil and the growth of crops. On Malta the temples became larger and more elaborate as time went on. This suggests that the ceremonies became more involved and that the priests became more and more important.

The Hypogeum of Hal Saflieni

The Hypogeum is a large complex of underground burial chambers built at three different levels. There are twenty separate chambers cut into the rock face. When archaeologists excavated the rooms in the early part of this century, they found thousands of human bones. They estimated that between 6,000 and 7,000 people were buried there.

The elaborate design of the Hypogeum suggests that ceremonies were performed there when people were buried. Some of the rooms have strange holes in the wall which go through to the room next door.

What were these?

One idea is that they were "oracle holes." Someone standing in the outer room would ask questions through the hole. The eerie, echoing voice of a priest hidden away in the inner room would reply.

The sleeping lady from Hal Saflieni.

The temples at Tarxien

The main picture shows the temple being built. The stone walls can still be seen today, but the thatched roofs are only guesswork. The plan below shows the rooms and passages which linked them. The rooms are of different sizes. The buildings are round or oval, but each is divided into a pair of semicircular chambers. The largest pair of chambers is at the front of the temple and the rooms get smaller as you go further into the temple. Perhaps the people gathered in the large outer chambers and the priests stayed in the inner rooms.

The insides of the buildings are decorated with leafy patterns, but there is no carving on the outside. The carvings become more elaborate in the later temples.

The people must have had sharp tools to make these carvings. Stone Age people made tools by striking a hard stone, such as flint, against another pebble to flake bits off and shape the end. There is plenty of stone on Malta, but no flint. We do not know if the people made tools from bone or animal antlers, or if they began to trade with other islands in the Mediterranean and imported supplies of flint.

Some of the stone slabs in the buildings look like altars, which is further evidence that they were temples.

The Earth Mother

Statues like the ones found on Malta are usually described as "fat lady" or "earth mother" figures. Many of these figures are very fat indeed. They are often sitting on a stone bench. But the puzzling thing about the Maltese statues is that you cannot tell whether they are male or female. However, in many early cultures these fat figures represent fertile soil and plentiful crops; if someone is fat and well fed, there must be plenty of food to eat. Most early peoples had at least one god who represented the fertility of the land. If the Maltese people did worship gods, it is very likely that they too would be connected with farming and crops.

Not all the statues are these fat figures. Some female statues have also been found. Two of these figures, which both come from Hal Saflieni, are lying on stone benches like the ones the fat figures sit on. One of these, known as the "sleeping lady," is naked to the waist. She is carved much more realistically than the fat figures (see page 15). Some naked standing female figures have also been found.

Who are these women meant to be? Are they goddesses and if so, how do they fit in with the fat "earth mother" figures? Once again, we do not know the answers. But the statues must have had some magical or religious significance. They were certainly not carved to look like the people themselves. We know from their bones

Left: The Venus of Willendorf.

Right: The Venus of Brassenpouy.

and carvings they left behind, that the early Maltese people were short with long heads. Their hair hung long and straight, though some people wore it in pigtails.

In about 2500 BC the temples were abandoned and the people disappeared. Perhaps their prayers to the Earth Mother and other gods had failed them. If there was a drought, for example, the crops would fail. For people who grew just enough to live on, this would be a disaster. There would not be any stores of food, so famine would wipe out most of the population.

A fat stone figure from Hagar Qim.

The first architects

How did the people manage to design and build such complex temples? The fact that the layouts are all similar suggests that someone must have had the original idea and then worked out the details very carefully.

Nowadays, an architect draws up plans for a new building. The early Maltese people did not draw as far as we know, but archaeologists have found small models of buildings made of limestone. So they must have had their own architects who designed the temples and made the models to show the builders what to do. The models also help us to see what the temples would have looked like.

After the architect had made his model design, the workers started building. There was plenty of stone but moving it into place was a massive task with no lifting equipment. How did they do it? One clue is the balls of limestone which were found near Tarxien. They could have been use to roll the large building blocks into place.

A piece of carving from a temple.

The building techniques and layouts of the temples improved as the workers became more skillful. The first temples were made with rubble and large stones were used only for the outer walls and the doorways. Later buildings, such as the temple at Tarxien, were built entirely from giant slabs of stone.

None of the roofs has survived, so we have to guess what they looked like. The main problem would be supporting the roof. It seems that the inner walls sloped in towards the middle of the building. This was done by laying each layer of stones slightly nearer the centre of the room. Sloping walls gave a narrower roof opening and it was possible to lay a beam and roofing material such as thatch across the gap.

Found by accident

No one knew about the temples on Malta or the unknown people who built them until building workers accidentally discovered the Hypogeum at Hal Saflieni in 1902. Even so, excavations did not begin until 1907 when the archaeologist Themistocles Zammit began work on the Hypogeum. He then carried out important excavations at Tarxien and the other Maltese temples.

The books he wrote about the temples aroused the interest of archaeologists elsewhere and a full study of all the ruins was carried out after the Second World War by the Royal University of Malta.

Knossos

Crete, c. 2000–1450 BC

Did King Minos keep a terrifying monster in the labyrinth at Knossos?
Were ancient Athenians sacrificed in the bull dance? And why did such stories begin
about the peaceful and artistic Minoan civilization?

Crete is the largest of the Greek islands and has been inhabited longer than any of the others. In fact, there is evidence that people were living there 9,000 years ago. By about 2500 BC the island had become the center of one of the greatest civilizations of the ancient world. The achievements of its people ranked with Egypt and Mesopotamia.

Who were they?

However, we do not know even what these people were called. In fact, the civilization was almost unknown until 1900, when archaeologists uncovered the ruins of the great palace at Knossos. Who were these people? Written records left no clues, but archaeologists had to put a name to the creators of the lavish palaces and treasures they found. So they decided to name them the Minoans after their legendary ruler, King Minos.

Mystery surrounds the Minoans and there are many questions we cannot answer. They seem to have arrived on Crete in about 7000 BC and they probably came from Asia Minor. It is not surprising that they chose Crete. It is a mountainous island, but the soil is good and there is more rainfall than in other parts of the Mediterranean, so the people could farm in the fertile valleys between the mountains. And because they were divided by mountains, the people formed separate communities.

The landscape of the island created natural barriers against enemies, so the people lived peacefully. By about 2000 BC they had begun to build their first palaces. Evidence shows that these early palaces were rebuilt and added to as time went by. For example, there are ruins of a much earlier building under the palace at Knossos.

Knossos is the most

Minoan women were treated as men's equals, which was very unusual in the ancient world. They had elaborate hairstyles and used makeup on their lips and eyes.

DID YOU KNOW?

These images appear again and again in the wall paintings at Crete.
The woman with a snake is either a priestess or a goddess, and she is riding a bull, which
was also worshiped as the symbol of Poseidon, the Greek god of the sea.

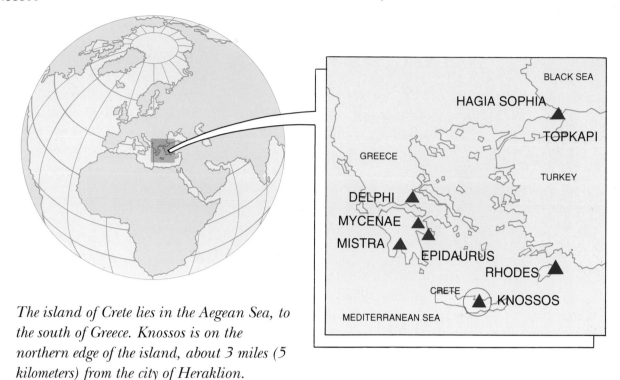

The island of Crete lies in the Aegean Sea, to the south of Greece. Knossos is on the northern edge of the island, about 3 miles (5 kilometers) from the city of Heraklion.

magnificent but it is not the only palace on Crete. There are four others – Phaistos, Mallia, Myrtos and Zakro. Each one was the center of a community, and the government of the people was carried out from there. But who lived in them? And why did they need such huge and elaborate buildings?

One theory is that the palaces were the centers of Minoan trade. At Knossos Sir Arthur Evans found dozens of huge jars, each one taller than a man. They were used to store enormous quantities of oil and wine. At Mallia there are eight large structures which were probably granaries. Perhaps grain, oil, and other produce were stored at the palaces for trading.

Sir Arthur Evans also found evidence that Knossos was used for religious ceremonies. He came to the conclusion that a "priest king" lived there and ruled over the people. A person who controlled both trade supplies and religion would become very powerful and very wealthy indeed.

Trading across the Mediterranean
The Minoans could not have built up their civilization without trade. Crete is within easy sailing distance of Egypt and various islands in the Mediterranean. Ships could also reach the important caravan routes through the Near East that crossed overland to Mesopotamia.

The fertile agricultural land meant that the Minoans could grow more food than they needed. This was probably the basis of their trade at first. But as they built up their wealth, they turned their attention to crafts such as pottery, jewelry and metalwork.

The Minoans had to import raw materials for their crafts. We know that Cretan ships sailed the Mediterranean, buying gold and pearls from Egypt, silver from the Aegean islands, ivory from Syria, and ostrich plumes from

North Africa. They also imported tin from Spain to make bronze.

They exported wine and olive oil, and also the luxury goods which they made. Examples of their exquisite pottery and metalwork have been found in Egypt, Greece, and Cyprus. The quality of their workmanship shows that the Minoans were much more sophisticated than other peoples in the ancient world.

Tributes for the palaces

A good many of the people must have been craftspeople. They made luxury items for the five Cretan palaces, as well as for trading. The craftspeople would have lived in the towns which grew up around the palaces. Country villas also had workshops for craftspeople, while some craftspeople worked from the palaces themselves.

Most of the other people would have been farmers who lived in small country communities. However, we do not really know how the people were organized. One theory is that there was a tribute system between the palaces and the working people. The palaces would take a large amount of the products or food produced by the workers as a sort of tax. The workers would be allowed to keep the rest to use in whatever way they wished. For example, extra food could be traded for pottery and other household items. Another possibility is that some workers, particularly those from the palaces, were slaves.

The occupants of the palaces, whoever they were, must have been surrounded by nobles who formed the upper classes of Minoan society. They enjoyed a lavish lifestyle, surrounded by beautiful things. There do not seem to have been any wars on Crete and life in the Minoan heyday was peaceful and happy.

Theseus and the Minotaur

The legend of King Minos came from the ancient Greeks. The story goes that he ruled his people like a tyrant from his palace at Knossos, which was designed for him by the architect Daedalus. At the centre of the palace was a labyrinth or maze where Minos kept a fearsome monster, the Minotaur, who was half man and half bull.

A bull's head in stone.

King Minos's son died in an accident while visiting Athens. The king agreed not to wage war on Athens only on the condition that each year they sent him seven young men and seven young girls to be fed to the Minotaur. Theseus, son of the King of Athens, volunteered to go as one of the victims so that he could kill the monster. At Knossos he met Minos's daughter, Ariadne, who fell in love with him. She gave him a spindle of wool to unwind as he went through the labyrinth, so that he could find his way out again. She also gave him a sword with which he killed the Minotaur before escaping from Crete with Ariadne.

23

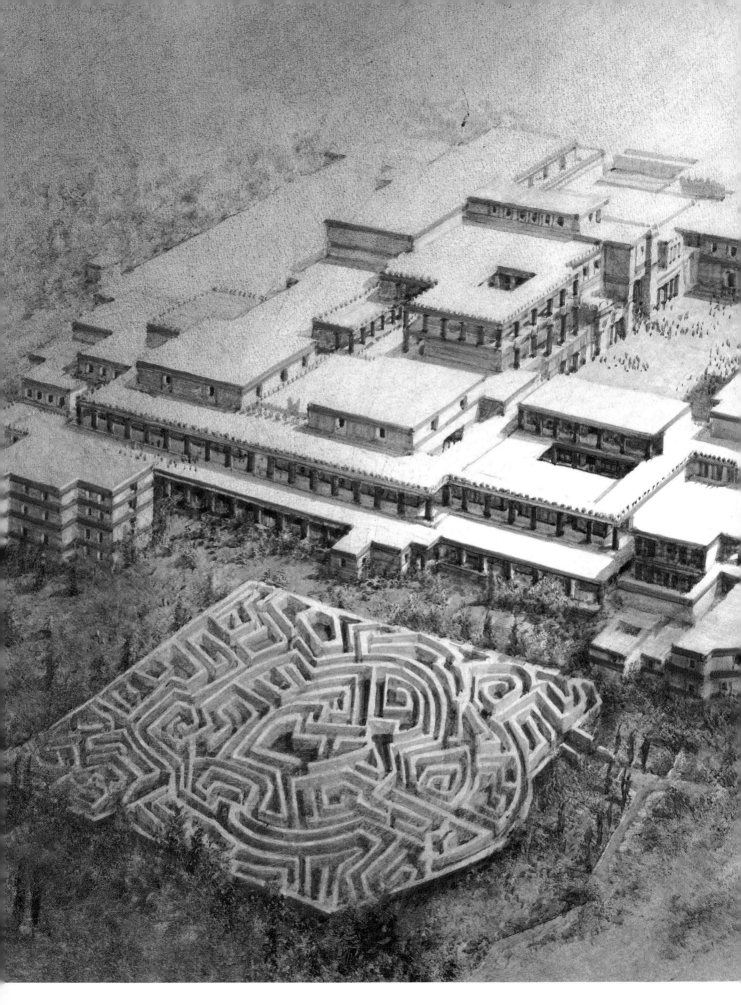

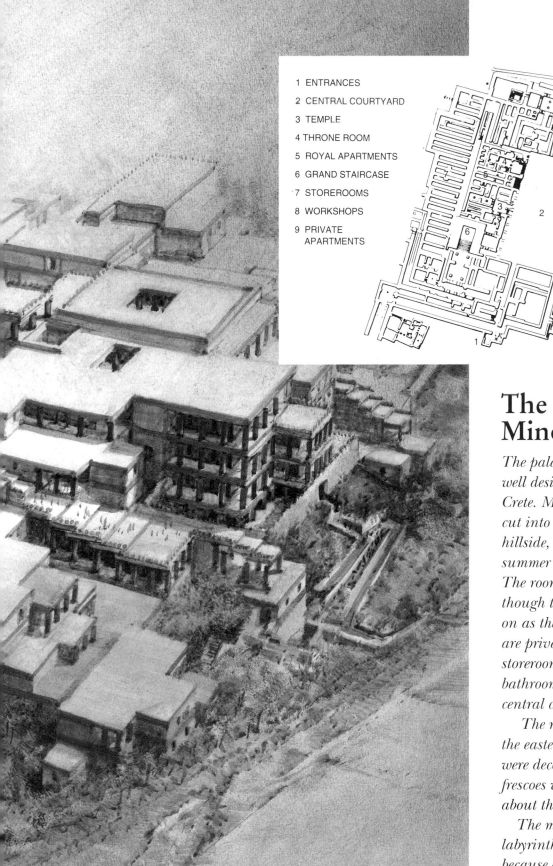

1 ENTRANCES
2 CENTRAL COURTYARD
3 TEMPLE
4 THRONE ROOM
5 ROYAL APARTMENTS
6 GRAND STAIRCASE
7 STOREROOMS
8 WORKSHOPS
9 PRIVATE APARTMENTS

The palace of Minos

The palace at Knossos was well designed for the climate of Crete. Many of the rooms are cut into the rock of the hillside, so they are cool in summer and warm in winter. The rooms themselves look as though they have been added on as they were needed. There are private apartments, halls, storerooms, workshops, and bathrooms, all built round a central courtyard.

The royal apartments, on the eastern side of the palace, were decorated with wonderful frescoes which tell us more about the Minoan way of life.

The myth about the labyrinth probably arose because of the maze of passages which connect the rooms. It would be easy enough to get lost in them. In fact, the palace itself may have been known as the labyrinth.

Religion and bull-leaping

The wall paintings at Knossos give us clues about Minoan religion. The religious symbol of the labrys or double axe appears in many rooms. There are also many paintings and statues of goddesses holding snakes. They always have bare breasts and may have represented a type of earth mother. It would be natural for the Minoans to worship an earth goddess because their life was centered around the land and the sea. Many of their paintings show plants and animals, such as birds and fish, which illustrates their love of the natural world.

Carvings found on Crete show that the Minoans sacrificed animals as part of their religious rituals. This was common practice in the religions of the ancient world. However, excavations at Knossos in 1979 revealed a more sinister side to their religion. In a building near the palace, archaeologists found a store of vessels and other objects which must have been used for religious rituals. With them were the bones of four children and part of a sheep's skeleton. The bones contained many knife marks. We cannot be sure what this means, but the fact that they were found with religious objects suggests that they were sacrificed. This is the only evidence of human sacrifice, if that is what it was. One theory is that the children were sacrificed to prevent some particularly feared disaster.

Many of the wall paintings show the ritual known as bull-leaping. The bull was worshiped as a symbol of power on many of the Mediterranean islands, but on Crete it had an extra significance.

The snake goddess.

The island often had earthquakes and the people believed that these were caused by the bull who lived under the island. When he was angry he shook himself so hard that he made the land erupt.

This led to the ceremony of bull-leaping, which was probably a religious ritual carried out to placate the earth-bull. It was performed by athletic young men and women and was very dangerous. The athlete would grasp the bull's horns and somersault over his back when he flung them into the air. These ceremonies may have been carried out in the palace courtyard, but it is more likely that they took place in special arenas outside the palaces.

Bull-leaping.

A sophisticated lifestyle

The Cretan palaces are spectacular but they are different from other monuments of the ancient world. They do not contain any fortifications, so the Minoans obviously did not feel the need to defend themselves from enemies. Their architects seem to have concentrated on beauty and comfortable living. But they were also way ahead of their time.

Some of the most spectacular features of Knossos are the huge staircases which are built out of large blocks of stone. Building these in ancient times was incredible enough, but there is another aspect which does not appear anywhere else in the ancient world. The Grand Staircase leading to the private apartments contains a drainage channel to carry away waste water from the upstairs rooms. The channel is cut in zigzags to slow down the water and stop it from overflowing.

Large towns grew up around the five palaces. Knossos was the biggest; at its peak it may have had a population of 100,000 people. This would have made it the largest European city of its day. The palace was built on the hillside and a viaduct led across a ravine to the town. The houses in the town, the homes of craftspeople and other workers, would have been quite small and simple.

Because of the earthquakes, the palaces and other buildings had to be

A model of a Minoan house.

Part of a fresco from Knossos.

rebuilt several times. Knossos was built right across a fault in the earth's crust and was destroyed and rebuilt at least twice. Each time, the palace was built on a grander scale. The walls were made of large mud bricks, but there was also a wooden framework and stout wooden columns to strengthen the building and help it to withstand the earthquakes.

Excavations at Knossos show that even early building work was quite sophisticated. The original settlement which was found under the palace ruins consisted of houses connected by roads paved with stone. But Minoan architects did not plan their buildings in a symmetrical way, like the architects of ancient Greece. In the palaces, for example, the rooms seem to have been added on haphazardly, without an overall design.

Two mysterious alphabets

When Sir Arthur Evans excavated Knossos, he found a number of clay tablets covered with a form of writing. There were three different scripts. The earliest script used hieroglyphs, a form of picture writing. The other two scripts were more complex. Symbols were still used but they were not simple pictures. Evans called these two scripts "Linear A" and "Linear B." To this day no one has managed to decipher Linear A.

Linguists have done rather better with the second script, Linear B. In 1939 tablets written in Linear B were found at Mycenae in Greece (see page 31). It became clear that

this was the script of the warlike Mycenaeans who dominated Crete after the Minoans. But no one managed to read it until 1952 when the British architect, Michael Ventris, came to the conclusion that it was a primitive form of the Greek alphabet. It was then possible to decipher what the tablets said. But it seems that the Minoans' Linear A will remain one of the mysteries of the ancient world.

A piece of Linear B script, giving a farmer's stock list, and a Minoan pot.

Master craftspeople

Excavation of the palaces suggests that different areas specialized in certain crafts. Bronze articles have been found at Phaistos and beautiful ivory work was produced at Zakro. Mallia was the home of the goldsmiths who produced breathtaking jewelry and luxury articles for the nobility. Knossos was famous for its jewelers who worked with precious stones, and for its seal carvers.

Minoan carved seals provide us with some of our most interesting information about the people – how they dressed, and the ceremonies and rituals they performed. The earliest seals were made from soft stone, but later craftspeople found ways to carve

harder rocks such as agate and jasper. The exquisitely detailed carvings on the seals showed animals, people and ceremonies such as bull-leaping.

Bronze-working was one of the most important Minoan crafts. As well as luxury items such as vases, the bronze-workers made tools like saws, axes, and pick-axes for other craftspeople to use.

Potters made vessels of all shapes and sizes, many of which were richly decorated with pictures of fruit, flowers, animals, or sea creatures. The Minoans also made models in faience, a form of glazed earthenware. Many were of their earth goddesses, but there were also models of plants and animals.

Controversial reconstructions

Most of what we know about the Minoans comes from the work of Sir Arthur Evans who began digging on Crete in 1900.

As well as uncovering the remains of Knossos, Evans restored some of the frescoes and parts of the palace such as the Grand Staircase. Many people have criticized his work, which was revolutionary at the time. But thanks to him, visitors to Knossos have a far clearer idea of what the palace would have looked like.

Other archaeologists have worked at Knossos and the other palaces since then, and more evidence has been uncovered, such as the bones which suggest human sacrifices.

The fall of the Minoans

What happened to the Minoans? They seem to have vanished into thin air in about 1450 BC. We know something about their disappearance but there are still unsolved mysteries.

At about this time the volcanic island of Thera exploded with such violence that rocks and volcanic ash were hurled for miles. It was probably one of the biggest eruptions ever known and, though Thera was 70 miles (110 kilometers) away, hundreds of tons of volcanic ash reached Crete and buried its cities and palaces. Archaeologists think that the eruption also caused a massive tidal wave or tsunami which smashed onto the island, demolishing everything in its path.

Yet there is another puzzle. Archaeologists found no evidence of people buried under volcanic ash, as happened in Pompeii when Vesuvius erupted. It was as if the Minoans had some warning of the coming disaster and managed to escape.

Whether the Minoans escaped or not, their time of prosperity was drawing to an end. The warlike Mycenaeans were beginning to dominate the Aegean. We do not know if they threatened the Minoans before the eruption of Thera, but they would certainly have waged war on them before long. The days of peaceful trading were over.

The volcanic eruption on Thera.

Mycenae

Greece, c. 1600–1100 BC

For centuries people have been intrigued by ancient Greek myths about a bloodthirsty people living in a grim hilltop fortress – tales of gods and kings, conspiracies, murder, and bloodshed. But Mycenae was not a mythical place at all...

For thousands of years people in ancient Greece told stories of a mysterious warrior people who had lived in a hilltop fortress. They were tales of gods and heroes, jealousy and murder. The Greek poet Homer described these people in his famous poems, the *Iliad* and the *Odyssey*. He called them the Achaeans.

But where was their fortress? There were no signs of the ruins anywhere. Were the stories just legends?

Then, in the late nineteenth century, a German businessman, Heinrich Schliemann, decided to find the lost fortress. He studied Homer's poems and followed clues contained in them, and he made some incredible finds. Not only did he discover a fortified city high on a desolate hilltop at Mycenae, but he also found graves cut into the rock. In these graves was one of the largest collections of gold treasures ever found. So who were these people who amassed such vast wealth. Were they Greeks? Where did they come from?

Warriors of the Mediterranean

The first settlers in Greece were a people called the Helladic. They were farmers who lived in stone houses and could make bronze tools, but they were a simple people compared with the Minoans.

Then suddenly, in about 1600 BC, there was a change. The people became wealthier and more skilful. They began to build larger towns and bury their dead in lavish graves filled with treasures. Why did this change come about?

One reason is that the Mycenaeans, as these people came to be called, traded with their neighbors and so made contact with the Minoan civilization on Crete.

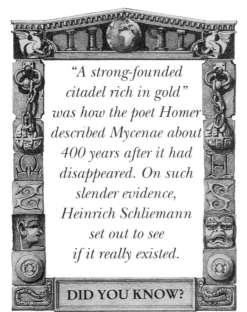

"A strong-founded citadel rich in gold" was how the poet Homer described Mycenae about 400 years after it had disappeared. On such slender evidence, Heinrich Schliemann set out to see if it really existed.

DID YOU KNOW?

The inside of the palace at Mycenae would have looked something like this, with decorated walls and ceiling. Behind the Mycenaean warrior-king is the beaten gold death mask which might have belonged to Agamemnon.

Mycenae

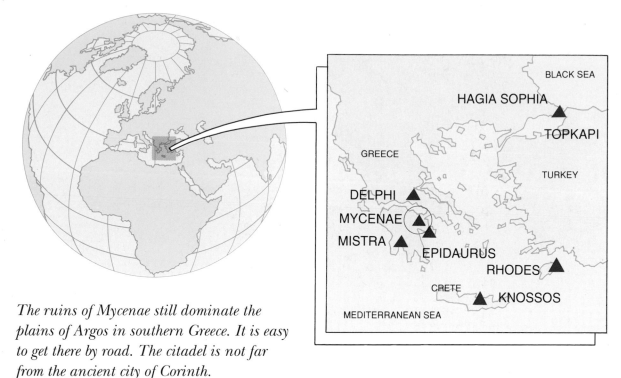

The ruins of Mycenae still dominate the plains of Argos in southern Greece. It is easy to get there by road. The citadel is not far from the ancient city of Corinth.

They borrowed many techniques from the Minoans, such as wall painting, decorating pottery, and making seals.

The Minoans were peaceful and artistic, while the Mycenaeans were warriors who were greedy for power. Before the Minoan civilization disappeared, the Mycenaeans were already raiding the Mediterranean lands in search of tribute, and they soon became rich and powerful. Mycenaean swords were found in the ruins of Knossos and this has led people to ask whether some of the gold at Mycenae was booty stolen from the Minoan palaces. However, the Mycenaeans were excellent craftspeople in their own right, as well as fierce warriors.

Tracing their roots
The Mycenaeans had

The Lion Gate, the main entrance to Mycenae.

disappeared by the time of the rise of the great civilization of classical Greece. Is there a link between them and the ancient Greeks, apart from the fact that they lived in Greece? Or were they a completely separate people?

One clue to their identity was provided by Michael Ventris when he discovered that the Linear B script (see page 28) on tablets found at Mycenae and on Crete was an early form of Greek. Tablets written in Linear B were also found in other parts of the Greek mainland.

Once Ventris and other scholars had managed to decipher the tablets, people learned a great deal more about the Mycenaeans. One tablet found at Pylos, another of the Mycenaean cities, tells of offerings made to the same gods and

32

goddesses as those worshiped in classical Greece. This showed that the Mycenaeans were Greeks in language and religion at least.

Life in Mycenae

The Linear B tablets and the grave treasures tell us a good deal about the Mycenaean way of life. There was obviously a sizeable population living at Mycenae, which was the center of the Mycenaean kingdom, and their society was well organized. The people were ruled by a king who lived in the palace. There were also war leaders, officials, and councilors who were the nobility.

Trades mentioned on the tablets include bakers, carpenters, bronze-workers, potters, masons, messengers, and heralds. Goldsmiths made luxury goods for the nobility, and armorers made armor and weapons. Swords, daggers, and spears have all been found. Some of the swords were very elaborate, with decorations in gold and silver.

One strange thing about Mycenae is that there are no temples, so we do not know where they worshiped a large number of gods, including the later Greek ones and the snake gods of the Minoans. Evidence has been found of animal sacrifices. But that is really all we know about Mycenaean religion.

What happened to the Mycenaeans?

Why did such a powerful and successful people disappear? The change in their fortunes came in about 1250 BC, when a series of fires swept through their land. There were three fires in all, and they wiped out all the Mycenaean cities.

We do not know how these fires started, but it seems that they were deliberate. There is no evidence of an invasion and no culture took over from the Mycenaeans, so the most likely explanation is that some of the Mycenaeans themselves destroyed their cities, perhaps because they were jealous of their wealthy rulers. And with the destruction of the cities, the Mycenaean culture began to decline.

A king's head in ivory.

The Treasury of Atreus

One of the most important buildings at Mycenae was known as the Treasury of Atreus. Atreus was the father of Agamemnon, the warrior-king of Homer's *Iliad*. In fact, the building is not a treasury but a tomb, though we do not know who was buried there. However, the tomb is so magnificent that it must have been for a king.

It is shaped like a huge beehive, about 50 feet (14.5 meters) across and is approached along a narrow corridor. The structure is made of stone blocks, and each one once contained a bronze decoration. The massive doorway would also have been decorated.

The tomb must once have been full of treasures, but many of these have been stolen. Many royal tombs contained masks of beaten gold, which were obviously meant to represent the king buried there.

Unconquered citadel

The palace at Mycenae sits on top of the hill and the whole citadel is surrounded by a massive wall. The main entrance to the city is the famous Lion Gate. The circle on the right of the picture is a graveyard; there are houses behind the graveyard and near the Lion Gate. Some of the houses, such as the House of Columns, were large stone structures which were obviously the homes of military leaders and nobility. Craftspeople and other workers probably lived in smaller houses within the walls.

The palace complex was large, but it did not have the elegance of the Cretan palaces, with their columns and balconies. The Mycenaeans seemed more interested in providing strong defenses. However, the inside of the palace was lavishly decorated with wall paintings.

1 LION GATE
2 STORE ROOMS
3 POSTERN GATE
4 GRANARY
5 GRAVE CIRCLE
6 PRIVATE HOUSES
7 WEST PORCH
8 PALACE
9 GREAT COURT
10 HOUSE OF COLUMNS

Agamemnon and Clytemnestra

Agamemnon was a king of Mycenae and the commander of the Greek army during the Trojan War. The legends surrounding him come to us from Homer's *Iliad* and from the *Oresteia,* a trilogy of plays by the Greek dramatist Aeschylus. They are full of violence.

The legends say that Atreus, king of Mycenae, had two sons, Agamemnon and Menelaus. When they were still infants, Atreus was killed and his throne was seized. The two boys were taken to the court of King Tyndareus of Sparta.

Menelaus married the beautiful Helen, who, the legend has it, was the daughter of Tyndareus's wife Leda and the god Zeus. Tyndareus gave up his throne and Menelaus became king of Sparta.

Helen had a sister, Clytemnestra, who was married to King Tantalus of Pisa. Agamemnon, who by now had won back the throne of Mycenae,

A bronze suit of armor.

declared war on Tantalus, killed him in battle, and forced Clytemnestra to marry him. And so the scene was set for the Trojan War.

Paris, one of the sons of King Priam of Troy, fell in love with Helen and took her back to Troy. Agamemnon and Menelaus set out with a large fleet of ships to rescue her, but it was to be ten years before they captured Troy, rescued Helen, and returned.

Meanwhile, Clytemnestra, left alone at Mycenae, was scheming. She hated Agamemnon because he had killed her first husband and forced her into marriage. While he was away, she took a lover, Aegisthus. Then news reached Clytemnestra that Agamemnon was returning triumphant from Troy with a new Trojan lover, Cassandra. Her bitterness and hatred grew even greater and she sought revenge. She and Aegisthus plotted to kill Agamemnon and Cassandra when they returned.

Clytemnestra greeted Agamemnon lovingly. She led him to the bathhouse where slaves had prepared a bath for him. But as he was getting out of the bath she entangled him in a net and Aegisthus stabbed him twice. Then Clytemnestra beheaded Agamemnon and went out to kill Cassandra. She seized power in Mycenae, which she ruled over with Aegisthus. Years later she and Aegisthus were killed by her son Orestes, who thus revenged his father's murder.

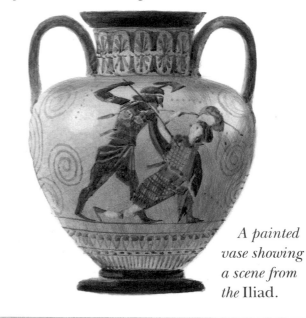

A painted vase showing a scene from the Iliad.

Wheeling and dealing

The Mycenaeans developed an organized system of trading and must have been rivals of the Minoans. At the height of their power, their ships carried goods to and from most of the Mediterranean countries. The Linear B tablets shed quite a bit of light on the way they traded. Mycenae was the center for trading, but the other great cities, such as Pylos, Thebes, and Tiryns would also have had craftspeople and tradespeople working within their walls.

The cities were surrounded by small towns and villages where people produced or collected the raw materials needed. Craftspeople and other workers turned the raw materials

A golden goblet known as Nestor's Cup.

into manufactured goods which could be traded. Food was also grown in the countryside and brought into the cities. Each area had a stated quota of grain. The trading centers were involved at each stage and took some of the profit.

The Mycenaeans kept careful records of all these transactions. Some Linear B tablets give amounts of grain with details of the land it was grown on. Another group lists the amounts of bronze which should be contributed by the different areas. There are tablets about herbs and spices produced. Some of these lists may be amounts which people had to pay in taxes to the officials in the city. They show that the administration was well organized.

Treasure hunter extraordinaire

A German businessman and amateur archaeologist called Heinrich Schliemann discovered Mycenae but the work he did there was haphazard and made later excavations more difficult. Schliemann was intent on uncovering the cities of Homer, and on finding gold – and he was not disappointed. In the royal tombs he found gold death masks of kings, splendid gold tiaras for their queens, gold cups and bowls, and many beautifully made weapons.

The early tombs were shaft graves, which were little more than pits dug into the ground. The more lavish chamber tombs, such as the Treasury of Atreus, are known as tholoi.

Later archaeologists, such as Alan Wace and George Caro, made more systematic and careful excavations of the sprawling ruins, and particularly studied the shaft graves and chamber tombs. Their work helped people to understand how the Mycenaeans organized their lives.

Delphi

Greece, c. 650 BC–AD 150

Could the oracle at Delphi really see into the future?
What was the secret behind the mysterious rituals that went on at this
sacred site, the "center of the world?"

The sanctuary of Apollo at Delphi was one of the most sacred sites of the ancient world. People came from far and wide to this complex of buildings dominated by a magnificent temple on the cliffs of Mount Parnassus in northern Greece. They came because they believed that the god Apollo could tell them about the future. Kings and princes, heroes of Greek myths and legends, consulted the famous oracle of Apollo about all manner of problems.

But why did this particular site become so famous? There were temples and shrines to the gods all over Greece, so why did the ancient Greeks feel so drawn to Delphi? What was an oracle and why did the ancient writers have so many stories to tell about the oracle at Delphi?

Delphi, temple of mystery and myth

The ancient Greeks believed in many gods, but the twelve most important ones were supposed to live on Mount Olympus in northeast Greece. Zeus was the chief god and Apollo was one of his sons. He was the god of light, thought, poetry, and music, a god who could look into the future.

Ancient writings contain several myths about Delphi and why it became the most important place for finding out about the future. One of the earliest stories is about Delphos, the hero of Delphi, who was supposed to make predictions from animals' innards. Parnassus, the hero the mountain is named after, foretold the future by watching flights of birds.

But it was predictions made by the oracle which drew people from all over Greece to learn their fate at the Temple of Apollo.

Who was the oracle?

The oracle was meant to be the voice of the god

The first temple at Delphi was made of wood and was burned down. The Greek dramatists Aeschylus and Euripides describe the splendid stone temple which was built in about 500 BC.

DID YOU KNOW?

A priestess of the oracle. In the early days of the sanctuary young girls
were chosen for their purity, but later on older women could also become priestesses.
The figure in the background is pouring a drink-offering or libation to Apollo.

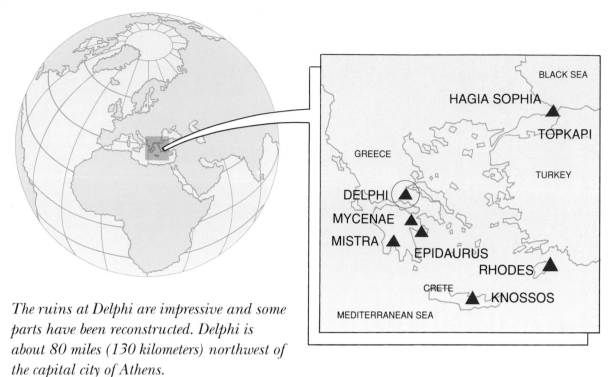

The ruins at Delphi are impressive and some parts have been reconstructed. Delphi is about 80 miles (130 kilometers) northwest of the capital city of Athens.

Apollo who spoke through a priestess called the Pythia. The earliest priestesses of the oracle were young women but later on older women were chosen as well. The Pythia wore the clothes of a young girl which was a sign that she would live a pure life and leave behind any family so that she could devote herself to the work of oracle.

What sort of questions did the oracle answer? Many of them seem to have been religious or political. The answers were usually fairly vague. For example, someone asking for the best way to worship might be told "according to the custom of the city," or "according to the gods."

Sometimes the replies were more precise. One story from the Greek geographer Strabo, tells of two men called Myscellus and Archais, from the city of Corinth, who went to

Delphi to ask the oracle where they should set up new colonies. The priestess asked them whether they wanted health or wealth. Archais chose wealth and was told to found Syracuse. Myscellus, who was a hunchback, chose health. The priestess told him his colony should be at Croton in southern Italy. Syracuse later became the richest colony in the west.

Some of the answers had more than one interpretation. One of the most famous stories is about King Croesus of Lydia, who asked the oracle if he should attack the Persians. He was told that if he did so, a great kingdom would be destroyed. But how could Croesus have known that the oracle meant his own kingdom and not the Persians!

Why Delphi?

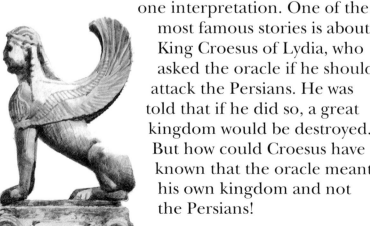

The Sphinx of the Naxians. Another myth raised the

importance of Delphi as a religious center for the ancient Greeks. Zeus wanted to find the center of the world, so he had two eagles released at exactly the same moment from the western and eastern ends of the earth. They flew towards each other at the same speed and met at Delphi.

In the fourth century BC the famous navel stone, called the omphalos, was set up at Delphi. It was flanked by two eagles to remind people that it was at the center of the world.

Three of these navel stones have been found at Delphi. This suggests that it is a very ancient site where early priests carried out unknown rituals and ceremonies.

Delphi was probably chosen because of its position. The area is very beautiful and rugged and was subject to earthquakes, which ancient people believed were caused by the gods. There is also a spring of fresh water.

Delphi was not just a religious center. Representatives from the city-states of Greece visited it regularly to consult the oracle, so it also became an important meeting place. Many of the city-states set up treasuries where they could make offerings to Apollo. These buildings were quite small but they contained great riches.

Ousted by Christianity
People still came to Delphi, even after the Greek empire had been taken over by the Romans. The priests of Apollo did not have the same power but the Romans still marvelled at the temple itself. When the Roman Empire became Christianized, however, the site lost its importance because the old gods and their rituals were discredited.

Consulting the oracle

In Delphi's early days the oracle could only be consulted once a year, but later there were divinations every month in spring, summer, and autumn.

The Pythia had a set routine. First a goat was sprinkled with cold water. If it shivered, Apollo was prepared to be consulted. The goat was sacrificed to Apollo and the Pythia carried out a cleansing ritual in the sacred stream. Then she sat on Apollo's three-legged throne behind a screen in an innermost corner of the temple.

The questioner would put his question to one of the priests, who took it to the Pythia. She fell into a trance and uttered strange words and cries. The priests translated these cries into verses, which they handed to the questioner.

The omphalos stone.

The sanctuary at Delphi

The sacred site of the sanctuary was in a walled enclosure on the hillside. The entrance was at the bottom of the hill. From here, a sacred way wound up past the treasuries of the Greek city-states to the temple of Apollo. The temple dominated the site. Its Doric columns and the huge statues which surrounded it could be seen all the way along the path. A statue of Apollo 50 feet (15 meters) high stood on the temple terrace. Behind the temple there was a theater.

A council of people, the Amphikytons, from all over Greece was responsible for looking after the sanctuary.

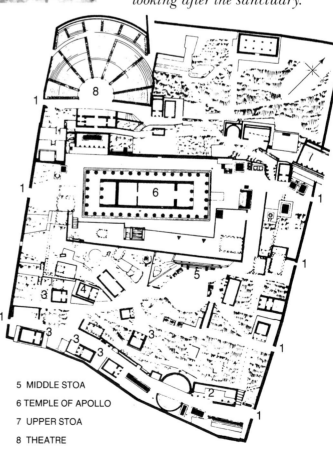

1 ENTRANCES

2 LOWER STOA

3 TREASURIES

4 ATHENIAN TREASURY

5 MIDDLE STOA

6 TEMPLE OF APOLLO

7 UPPER STOA

8 THEATRE

Wildness and wisdom

When people asked the oracle how they should worship, the reply was often "according to the gods." This meant the gods of Mount Olympus, particularly Apollo, and also the god Dionysus. This is strange, for Dionysus was the god of wine, quite the opposite of the gentle Apollo. Yet during the winter months, when the oracle could not be consulted, Dionysus was worshiped at Delphi.

A statue of Apollo.

According to Greek mythology, both gods were the sons of Zeus. Apollo was born to a mortal woman called Leto. Zeus's wife, Hera, was jealous of Leto and sent the she-dragon Python to pursue her so that she had to wander all over the world.

Gods do not take so long to grow up as mortals, and four days after he was born Apollo took a bow and arrow and set out to find the Python. He killed her and took over the Oracle of the Earth at Delphi. Apollo's priestess at Delphi was called the Pythia or Pythoness.

Dionysus, who was also called Bacchus, was the son of Zeus and the mortal woman Semele, daughter of King Cadmus of Thebes. The story goes that Zeus visited Semele disguised as a mortal. Hera, learning that Semele was expecting a child, was furiously jealous. She visited Semele disguised as an old woman and persuaded her to ask Zeus to appear in his true form. Unwillingly, Zeus did this, whereupon Semele was struck by a bolt of lightning and died. Her unborn child was sewn into Zeus's thigh. The child, Dionysus, was born three months later.

Dionysus is portrayed as a handsome but wild god and he was worshiped by bouts of wine drinking which resulted in frenzied behavior. His cult was very popular, perhaps because people enjoyed these wild festivals as a change from the more serious worship of other gods.

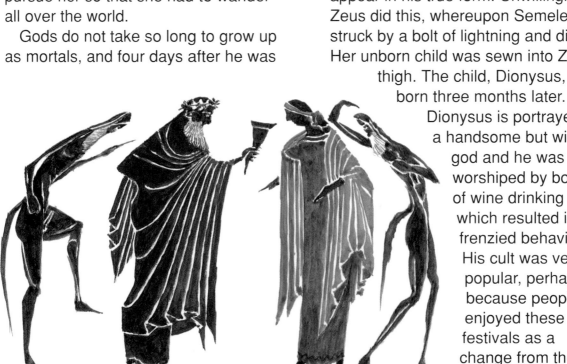

A vase painting of the rites of Dionysus.

Treasure troves

Athens, Corinth, Thebes, and many other Greek city-states had treasuries in the sanctuary, where they made lavish offerings to the god Apollo. The treasuries lined the sacred way and looked rather like small temples. They were magnificent buildings. Many contained beautiful statues and carvings. In fact, much of the Greek sculpture which you can see in Greece today comes from Delphi.

A carving from the Siphnian Treasury.

The Siphnians came from the island of Siphnos, which was one of the richest Greek islands. Their treasury contained marvelous friezes showing scenes from the Trojan War, and battles of gods and giants.

The Treasury of the Athenians was the first Doric building to be built completely of marble. Most of the treasuries are in ruins today, but this one has been restored so that visitors can see what it looked like.

The Pythian Games

People flocked to Delphi for another reason besides consulting the oracle – the Pythian Games. These games started as part of the festival to Apollo and included singing, recitation, drama, and music. The idea of drama as we know it today began in Greece as a way of honoring the gods. In the sixth century BC the games were reorganized by the Amphiktyons. Athletic competitions were also seen as a way of worshiping the gods with contests of strength. At Delphi events such as racing on foot in the stadium above the sanctuary or chariot racing on the plains below were added to the games.

The games became almost as important as the Olympic Games, which started at Olympia in Greece in 776 BC. Delphi's most famous piece of sculpture was inspired by the Pythian Games. This magnificent bronze statue originally showed a young charioteer driving a four-horse chariot. Today, only the figure itself remains but it is one of the most beautiful pieces of sculpture in the world.

The Charioteer.

Epidaurus

Greece, c. 320 BC–AD 150

Here, in the big open-air theater, actors of long ago performed some of the world's first plays. Here, too, sick people came to be healed by Asklepios. Were these two events linked? Who was Asklepios and how did he heal the sick?

If you go to the theater at Epidaurus, you can sit on the same seats and watch the same plays as the Greeks did over two thousand years ago. You can imagine the actors of long ago performing the plays of Sophocles and Aristophanes to an audience of 14,000 people who had come from far and wide to see them.

The theater was so large that the audience could not see the actors' faces, so the actors wore masks – sad for a tragic character and happy for a comic character. The masks of comedy and tragedy are still the symbol of theater today.

The theater at Epidaurus is the best preserved of any theater from the ancient world. Why was it built in this small town 31 miles (50 kilometers) away from the capital city of Athens? What was the attraction of Epidaurus that made people flock to it from miles around?

People were drawn to Epidaurus for another reason besides the theater. It also had a sanctuary to the god of healing, Asklepios. We are not sure how the cult of worshiping this god began. His name is surrounded by myths, but he was not one of the twelve major gods of Olympus.

Man or myth?

Epidaurus seems to have become a religious center in the eighth century BC. The people worshiped Meleatas, a local hero whose cult eventually joined with that of Apollo. Asklepios was thought to be the son of Apollo and a mortal woman, Coronis. As a result he was a mortal but with godlike qualities that enabled him to perform miraculous deeds on earth. We do not know whether the story of Asklepios is a complete myth. It may have be that there really was a man

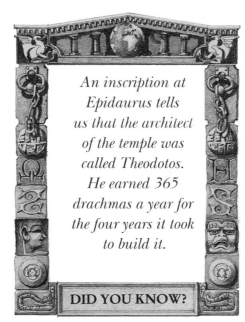

An inscription at Epidaurus tells us that the architect of the temple was called Theodotos. He earned 365 drachmas a year for the four years it took to build it.

DID YOU KNOW?

The theatrical masks of tragedy had a variety of expressions, but all had large, gaping mouths so the actors could be heard. Here the muse of tragedy, goddess of the theater, is flanked by sinister cloaked figures.

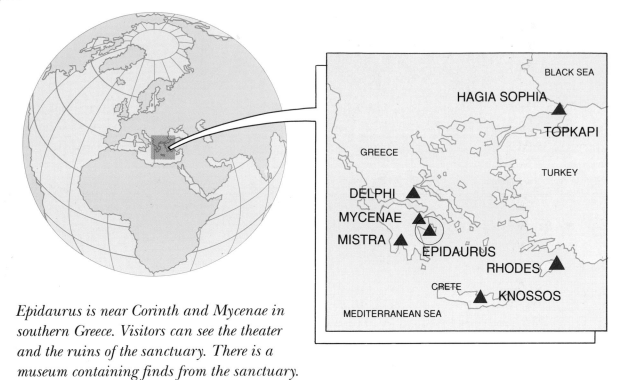

Epidaurus is near Corinth and Mycenae in southern Greece. Visitors can see the theater and the ruins of the sanctuary. There is a museum containing finds from the sanctuary.

who could heal people and who was made into a god after his death.

One myth said that Asklepios had been taught how to heal by a snake, so the symbol of Asklepios was the snake. The Greeks believed that snakes had a special knowledge of plants, especially medicinal herbs. Asklepios is reported to have healed supposedly incurable diseases. People believed that blindness could be cured by the snake of Asklepios licking the eyes.

Worshiping the healer

By the beginning of the fifth century BC an elaborate festival was held each year to worship Asklepios. Every spring there was a festival for Poseidon, the god of the sea, on the Isthmus of Corinth, the neck of land which links northern and southern Greece. The festival for Asklepios was timed to come just after this, in late April or early May, so that people could come straight on from Corinth to Epidaurus, which was

only 19 miles (30 kilometers) away. So once a year people flocked in to Epidaurus to mingle with those who had come to have illnesses cured.

We have to guess exactly what form the festival took. Animals were probably sacrificed and there would have been large banquets. We do know that there were athletic contests, such as wrestling, and artistic contests, which probably included singing, dancing, and drama.

A temple to Asklepios

By the fourth century BC, the cult had become so popular that a large and magnificent temple dedicated to Asklepios was built in the sacred area or sanctuary. It was built mostly from the local limestone, with columns made of Corinthian stone and a roof of terra-cotta tiles. Inside the temple was a large statue showing Asklepios sitting on a throne with a dog at his feet.

The sanctuary also contained a hostel where people could stay and a building called the abaton. This was for the people who came to be healed. When they arrived, they would go through a cleansing ritual and then spend the night in the abaton. A wall carving shows what is supposed to have happened next. Asklepios appeared to the sick people in a dream and suggested a cure for their ills.

In the morning they repeated their dreams to the priest who then treated the sick people with secret remedies. When people were cured, their names and details of the cure were carved on stone blocks. The priests of Asklepios were the only doctors, until Hippocrates introduced the basics of medicine.

There is another strange place in the sanctuary. It is a circular building called the thymele, which is surrounded by columns and contains a stone maze. We do not know what it was used for. Perhaps the sacred snakes of Asklepios were kept here, or it may have been the burial place of the healer. Another possibility is that it was a place where sacrifices were made to the god.

Preserved by a landslide
The cult of Asklepios declined with the coming of Christianity and the sanctuary fell into disrepair. The theater survived because something – perhaps a landslide – filled it up with earth which protected its stones. So today we have a near-perfect example of an ancient Greek open-air theater, where modern actors can recreate the dramas of classical Greece.

Godlike doctor

In Greek legend, Asklepios was the son of Apollo and Coronis, the daughter of Phlegyas, King of the Lapiths. There are different stories about him.

One says that Apollo left a snow-white crow to guard Coronis while he was away. Coronis was unfaithful to Apollo and the crow flew off to tell him. But Apollo had already sensed Coronis' unfaithfulness and cursed the crow for not pecking out her new lover's eyes. The curse turned the crow black, and all crows have been that color ever since.

This legend says that after Apollo's sister, Artemis, had killed Coronis with her silver arrows, Asklepios was brought up by Cherion the Centaur, who taught him healing. He was so skilled that he could raise people from the dead. Hades, the god of the underworld, complained to Zeus that Asklepios was taking all his subjects, and Zeus killed Asklepios with a thunderbolt.

Asklepios is always shown with snakes.

Greek drama

The theater at Epidaurus lies just outside the main area of the sanctuary, in a natural bowl made by the surrounding hills. It is a semi-circular, open-air theater of the kind found all over Greece.

The theater could have been designed by the architect Polykleitos in about 360 BC, but some scholars think it was built later than this because it seems to have been influenced by the theater of

Dionysus in Athens. It consists of three main parts: the theatron, which was the large, fan-shaped area of tiered seats for the audience; the orchestra, which was the circular, flat area in front where the plays took place; and the skene, a permanent background with a building behind where the actors prepared for their entrances.

The shape of the theatron meant that everyone in the audience would be able to see and hear

well. There were staircases all round the theater and a horizontal passage about two-thirds of the way up, so that the audience could get to their seats easily. The seats were made of wood resting on the stone tiers. Important people such as priests and officials sat at the front on seats with stone backs.

The word "orchestra" means dancing place, and the ritual singing and dancing took place here as well as the plays. An altar at the center front of the orchestra is a reminder that dance and drama were originally acts of worship.

The skene was probably built of wood and had dressing rooms and a place to store stage props and scenery. There were several doors in the skene through which actors could enter and exit. Movable scenery could be propped against the front wall and changed during the play.

A dramatic experience

What would it have been like to go to the theater at Epidaurus? People would have flocked there in their thousands and stayed for the whole day to watch several plays. They took their own food to eat in the intervals. What sort of performances would they have watched?

The earliest forms of theater were festivals of song and dance, which were performed in Athens in honor of Dionysus, the god of wine. A chorus of people danced and sang. Then speaking parts were introduced. At first, one or two actors stepped out of the chorus and spoke the lines. Later, the actors and the chorus were separate, but the chorus was still an important part of the performance.

Vase paintings show us what the chorus would have looked like. One vase in the Antikenmuseum in Basle, Switzerland, shows a tragic chorus of around 500 to 490 BC. They wear short tunics and tragedy masks. The masks have hair and cover the whole head, not just the face. The chorus was accompanied by a musician playing the aulos or double flute.

The actors looked different from the chorus. Both tragedy and comedy actors were heavily disguised. A tragic actor wore a mask which, in the early days at least, probably covered the whole head. It had the heavy, stylized features we still associate with tragedy masks – pronounced eyebrows, cutout eyes and a downturned mouth. The actor wore long-sleeved garments and kothornoi, which were boots that covered the legs.

Comedy actors looked quite different. They wore the stylized comedy masks with the upturned mouth, but in the early days of comedy, the costumes were meant to look like false skin and were quite grotesque. There were also choruses of creatures, such as birds, frogs, and wasps, to make the scenes more colorful. In later comedies the actors wore more conventional garments.

A mask for a tragedy.

A plan of the theater. The skene is at the bottom of the picture.

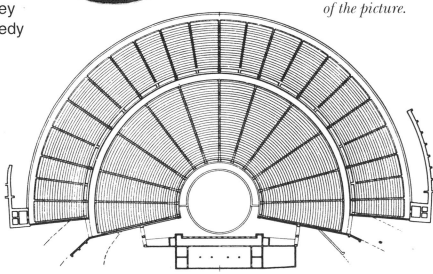

Playing to the gallery

By the classical period in Greece, plays told a definite story. There were three types of play. Tragedies were dramatic and solemn. Their theme was often the gods or heroes, so everyone knew the story already. The three most famous Greek dramatists were Sophocles, Aeschylus, and Euripides. They wrote more than 300 plays between them, but we have only thirty-three today.

Comedies were about everyday events and give us an insight into life in ancient Greece. Aristophanes was a writer of the Old Comedy period (c. 450–388 BC). Eleven of his plays survive today. One of them, *Lysistrata,* written in 412 BC, tells of the Peloponnesian war between Athens and Sparta, but in a bawdy way which was the style at the time. Menander's plays illustrate the New Comedy period (342–292 BC). They were more dignified stories about well-to-do people.

The third type of play were the satires, which made fun of serious stories and legends. A day's

A member of the chorus dressed as a bird.

program at the theater normally included three tragedies or three comedies, and a satire.

Excitement was added by the fact that the plays were in competition with one another. Dramatists would apply to an official for permission to enter a play for the festival. The priests and officials judged the plays and awarded a prize to the winner.

Grand reopening

Epidaurus was first excavated in 1881 by Panayotis Kavvadias and an archaeological team from Athens. Two German scholars, Armin von Gerkan and Wolfgang Muller-Wiener, have studied the theater in great detail, but can only guess at what exactly happened at Epidaurus. This is because all the play texts we have come from Athens and would have been performed in the theatre of Dionysus there.

By the time Epidaurus was built, comedies would have moved on from the Old to the New style. The tragedies would have changed from the formal style of Aeschylus to the more natural writing of Euripides. Aeschylus was an innovator and the first writer to introduce a second actor rather than using the chorus, but his plays still have a feeling of following a religious ritual which is missing from those of Euripides.

Rhodes

Greece, c. 280–226 BC

*The giant Colossus, the most mysterious of the Seven Wonders
of the Ancient World, towered over the city of Rhodes. But why was it there?
And what happened to it?*

Most people have heard of the Colossus of Rhodes. It was one of the Seven Wonders of the Ancient World, and in many ways it was the most fascinating. This massive statue, its huge legs astride the harbor entrance at Rhodes, has inspired writers through the ages. It stands as a symbol of power. In Shakespeare's play, *Julius Caesar,* Cassius uses the image to describe Caesar's might: "Why, man, he doth bestride the narrow world like a Colossus."

However, like many mysteries of the ancient world, the picture we have of the Colossus comes mainly from people's imagination. Famous though it is, we do not know what it looked like, exactly where it stood, or how it was built. Its life was short, for an earthquake brought it tumbling down only fifty years after it was built. Our only evidence is the word of a few ancient writers who saw it.

But we do know that it really existed.

So what was the Colossus? Why was it built? And how much of the evidence is fact rather than fiction?

City of power

Rhodes is an island in the northern Mediterranean. Dorian Greeks settled there and formed three separate city-states. At first these cities joined with the city-states of Greece as part of the Athenian league. However, they wanted to be independent of the powerful Athenians and in 407 BC they broke away and formed a new state of Rhodes. Their capital, also called Rhodes, was in the north of the island.

The island was well placed for trading with other Mediterranean powers and the Rhodians became rich and successful. The city of Rhodes was a symbol of their wealth. More than 60,000 people lived in the city, which was

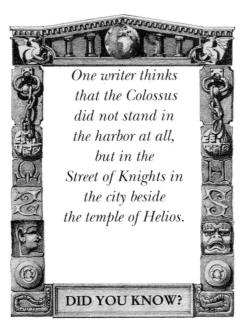

One writer thinks that the Colossus did not stand in the harbor at all, but in the Street of Knights in the city beside the temple of Helios.

DID YOU KNOW?

*No one knows what the Colossus looked like. It might have been
a classical statue or a strange-looking figure like this. But its sheer size would have meant
that it needed scaffolding to support it while it was being built.*

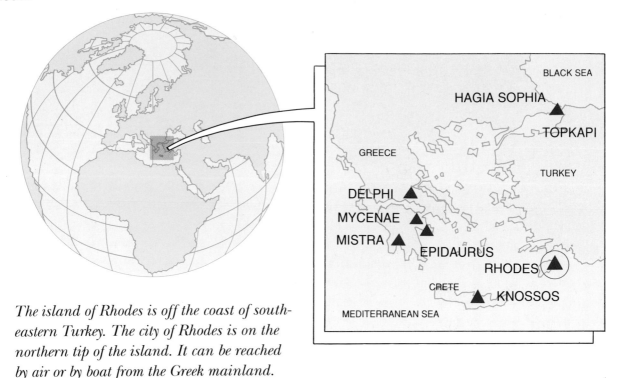

The island of Rhodes is off the coast of south-eastern Turkey. The city of Rhodes is on the northern tip of the island. It can be reached by air or by boat from the Greek mainland.

built in the classical Greek style. It had a well-protected coastline with a series of sheltered harbors for their trading ships. They also had a military fleet which protected trading vessels from Mediterranean pirates.

Victorious in war
Successful though they were, the Rhodians could not always remain independent. First they were conquered by Mausolus, king of Caria. Then Rhodes became part of the Persian Empire until Alexander the Great brought the island back under Greek control in 340 BC.

After Alexander's death, Rhodes became independent again, but kept its trading links with the new capital of the Greek empire, Alexandria in Egypt. They became allies of the Alexandrian king, Ptolemy I, and won a battle against Antigonus of Macedonia. But trouble was to follow. Antigonus asked the Rhodians to switch sides and join

him in his battle against the Egyptians. When they refused, he sent his son, Demetrius, to invade their island.

Demetrius descended on Rhodes with an army of about 40,000 soldiers and several siege engines. He was confident of success, but he had not bargained on the Rhodians' powers of resistance. They managed to fight off Demetrius and assert their independence.

Then they made an alliance with Antigonus against everyone except the Egyptians. It was to celebrate this victory that they built the Colossus.

Inspired by the sun god
The Rhodians chose Helios, the Greek sun god, as the subject for their statue. This may seem strange because many other gods were far more important in Greek mythology. However, Helios was the patron god of the three Rhodian states, Lindos, Ialysos and Kameiros – in fact, they are named after Helios's three surviving sons (see page 59).

A colossal undertaking

Ancient writers say that the Colossus was 110 feet (33 meters) high and that it took about twelve years to build. Other than that, we do not know much about what it looked like, but Philo of Byzantium described how it was put together.

The sculptor was Chares of Lindos and he made the statue in bronze. He achieved this by casting sections individually and then bolting them together. First a platform of marble was erected as a base for the statue.

A Greek helmet.

Then the feet were put in position, followed by the legs and so on. As the statue rose up, it must have been supported, probably by scaffolding.

The statue was reinforced by putting blocks of stone and diagonal iron struts inside. The extra weight would make it more stable and the struts would strengthen the structure. This was a new method, so Chares must have have been a skilled inventor and engineer, as well as a sculptor.

Brought to its knees

The geographer Strabo tells us about the disastrous earthquake which brought the Colossus down in 226 BC. He does not say exactly where the statue was situated, and its location is open to debate, but we know that the violent tremors of the earthquake made the statue break at the knees. According to one writer, many houses were crushed by the massive structure as it toppled onto the city.

The Rhodians left the three sections – the thighs, the body, and the head – that the statue broke into, lying where they had fallen. The ruins remained there for years to come. Why did they do this?

Strabo tells us that the people consulted an oracle who said they must not try to rebuild the statue. Perhaps that is why they left it where it was – to remind people of the city's greatness. When Rhodes became part of the Roman empire, the city lost much of its power, so the ruins of the Colossus would recall the days of the city's proud independence. Later generations did try to rebuild it, but failed.

A Greek ship.

The harbor of Rhodes

The harbor entrance would have been the natural place for the Colossus to stand. Here it would have been the first sight of Rhodes for visitors approaching by ship.

However, the romantic image of its great legs astride the harbor entrance just cannot be true. The entrance was about 1,300 feet (400 meters) wide and the bronze-workers could not have produced a structure strong enough to span

such a gap. It is more likely that the Colossus stood with its feet together, perhaps on one of the jetties of the harbor, or on a specially built island. Perhaps one arm was uplifted, holding a torch as a symbol of the sun. But we do not really know.

The city of Rhodes held a festival of games in honor of Helios, which was attended by representatives from city-states all over Greece.

Recorders of the past

Two writers of the ancient world give us much of the evidence we have about the Colossus of Rhodes, although neither of them can have seen it for themselves.

Pliny the Elder (c. AD 23–79) was a Roman scholar whose famous *Natural History* supplies us with many of the details we have about the classical world. It is a huge encyclopedia, of which only thirty-one books survive today. Pliny's eye for detail was amazing, but his work includes myth and folklore as well as fact. He died while investigating the eruption of the volcano Vesuvius which buried Pompeii.

Strabo (c. 64 BC–AD 21) was a Greek geographer who wrote seventeen books. He collected the evidence of other writers, as well as making his own observations, and his work gives us a great deal of valuable information about the ancient world.

Driving the sun chariot

According to Greek mythology, Helios, the sun god, was the son of Euryphaessa, the moon goddess. He is usually represented as a charioteer driving the sun across the sky each day.

One legend about him tells how his son, Phaethon, was always begging to be allowed to drive the sun-chariot. His doting mother, Rhode, tried to persuade Helios to agree and one morning he gave in and granted Phaethon permission.

Phaethon, who wanted to impress his sisters, set off in fine style, but he was not strong enough to control the white horses which pulled the chariot across the sky. First he drove them so high that everyone on earth shivered. Then he drove them so low that the heat of the sun scorched the fields.

The great god Zeus was very angry at this recklessness and so he killed Phaethon with a thunderbolt. He fell into the river Po and his weeping sisters were changed into poplar trees (some versions say alder trees) along the river bank.

Helios, the sun god.

Leptis Magna

Libya, c. 46 BC–AD 211

*Why did the Romans build this splendid city on the edge of the arid Sahara Desert?
And why was Leptis Magna, with its wonderful classical buildings,
abandoned to the encroaching desert sands?*

In north Africa, near the vast, arid wastes of the Sahara Desert, lay a magnificent Roman city. Leptis Magna, as the city was called, was one of the jewels of the Roman empire.

North Africa was one of the most successful parts of the Roman empire. The Romans planted grain in its fields, and built roads. Farming the fertile valleys of what is now Algeria brought them great wealth, and this is seen in the cities with their fine public buildings, temples, and baths.

But Leptis Magna was different. There were no great farms, because the desert was too close. Yet the Romans must have spent much wealth in constructing the city, with its ornate temples and superb sculptures. Why did they build such a fine city in the middle of nowhere?

A city is born
The importance of Leptis Magna was its position on the Mediterranean coast. There had been a city there long before the Romans arrived. The Phoenicians, who came from the eastern shores of the Mediterranean, had used it as a trading post in the seventh century BC.

The Phoenicians were a seafaring people. Their main trading port in north Africa was Carthage, to the north in what is now Tunisia. However, the Phoenicians wanted another base in north Africa and Leptis was ideal because it had a sheltered inlet at the mouth of the river Lebda, where ships could dock.

Leptis never became as important as Carthage but the Phoenicians built a sizable city there. We do not know much about what it looked like because the Romans did so much rebuilding, but it would have had a harbor and warehouses where trading goods would have been stored.

The Romans conquered Carthage in 146 BC.

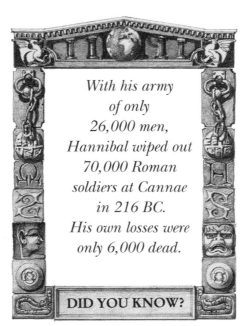

With his army of only 26,000 men, Hannibal wiped out 70,000 Roman soldiers at Cannae in 216 BC. His own losses were only 6,000 dead.

DID YOU KNOW?

*The temple, on its high platform, dominated the forum in Leptis Magna.
Here, Roman legionaries admire its classical lines, while the two-faced god Janus, patron of
new ventures and of the year's end, broods above.*

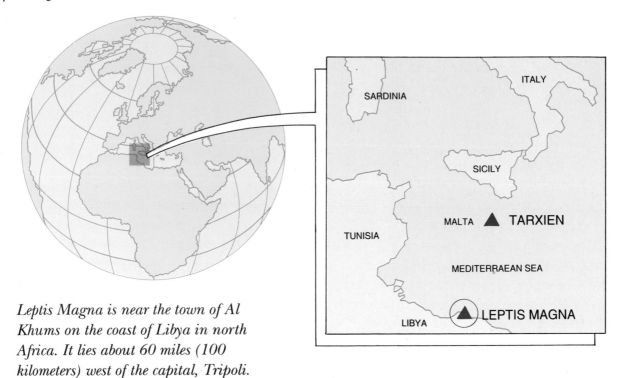

Leptis Magna is near the town of Al Khums on the coast of Libya in north Africa. It lies about 60 miles (100 kilometers) west of the capital, Tripoli.

This was the start of their domination in north Africa, though Leptis was not affected immediately. The city was brought into the Roman empire by Julius Caesar in 46 BC. The Roman rebuilding caused it to be renamed Leptis Magna – Leptis the Great.

Two cultures combine

Most of the people who lived in Leptis were Phoenicians who had come from the Middle East or from Carthage to settle there. But now their society was organized along Roman lines. It was divided into social classes – slaves, workers, and middle classes who lived in the city, plus a small number of wealthy landowners who formed the upper class. Centers of social life were the forum and the baths, as they were in Rome, and popular spectator sports included chariot racing and gladiator fights.

However, the Phoenician influence could still be seen. The gods had Roman names but they were the gods worshiped by the Phoenicians. Many of the people had Phoenician names, though some linked them with Roman names to make strange combinations such as Annobal Rufus Tabahpi.

Annobal Rufus was one of the most influential citizens of Leptis Magna in the early days of Roman occupation. He sponsored the building of the covered market and the theater.

The emperor from Leptis

Many of Leptis Magna's buildings were commissioned by the Roman emperor Septimus Severus. Severus was born in Leptis into an influential family, but though they were rich and successful, the family thought of themselves as north African rather than Roman. Some of them never even left Leptis or learned Latin.

Severus was different. He traveled widely and became very involved with the Roman empire. He worked for the emperor Marcus Aurelius in Rome, and then served in Athens for a while. Next he became governor of a province called Upper Panonia (in modern-day Austria) before becoming emperor in AD 193.

Severus was trained as a soldier and conducted many military campaigns to defend the empire during his reign as emperor. He campaigned in Asia, Egypt, and Nubia and went to Britain in AD 211, where he died at York.

There is nothing to suggest that he ever went back to Leptis after he became emperor, but Severus did not forget his birthplace. During his reign the harbor was extended and many magnificent new buildings were erected.

Abandoned to the sand

Leptis had always been a major trading post

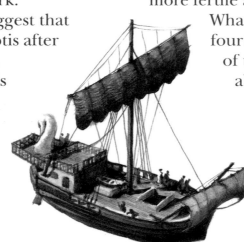

A Roman trading ship.

because of its position. Luxury goods, such as precious metals and stones, ivory, ebony, slaves, and wild animals, could all be shipped from its large harbor. But after the reign of Severus, the city seemed to decline.

One disadvantage had always been the lack of farmland. The Romans may have decided to combine trading luxury goods with shipping grain from the more fertile areas of north Africa. Whatever the reasons, by the fourth century AD many parts of the city had been abandoned. It was later invaded several times by barbarians, who tore down its walls. The number of people living in the city gradually dwindled, until finally it was deserted and left to the encroaching desert sands.

Carthage

The Punic Wars between Rome and Carthage began in 264 BC. The Romans won the first war in 241 BC, but the Carthaginians wanted their revenge. The great Carthaginian general, Hannibal, came up with a daring plan; the Romans controlled the sea routes across the Mediterranean, but they would not expect a land attack from the north. So Hannibal set out with 100,000 soldiers and 36 elephants to attack Italy via Spain and France.

The army had to cross the Alps in winter, and thousands of men died on the hazardous journey. When Hannibal reached Italy, there were only 26,000 left. Yet he and his army still managed to defeat the Romans in 218 BC and again in 217 and 216 BC. However, the Carthaginians could not hold out against the might of Rome forever and they finally came under Roman control in 146 BC.

Hannibal, the great general of Carthage.

The city of Leptis Magna

By the end of the reign of Severus, Leptis was a classical Roman city. The forum (a large paved area where people could meet and talk) was in the northeastern part, near the sea. The main street ran southwest from the forum. The buildings on this main street included the marketplace, the theater and temples. These were built in the period between c. 9 BC and AD 12, when Leptis was establishing itself as a Roman city. Smaller streets led off the main street at right angles in the grid system which was typical of Roman cities.

The market consisted of a long, rectangular courtyard with two octagonal buildings called pavilions. There were pillars around the courtyard and the buildings. Between these pillars and around the pavilions there were marble tables which served as market stalls. You can imagine a colorful market scene there, with the stalls laden with a variety of produce and people bustling about, buying and selling their wares.

The harbor had two sand-bars which curved into the sea from the river's mouth, and warehouses were built along these. There was also a lighthouse at the outer end of the eastern bar. Ships could dock in the shelter of the harbor while goods were loaded and unloaded.

Sweating it out

The public baths in Leptis were built during the reign of the Emperor Hadrian (AD 117–138), and so they were known as the Hadrianic Baths. They had hot and cold rooms, an open-air swimming pool, changing rooms, and lavatories. Bathers would go into the warm baths first, and then into the hot bath which was a room filled with steam. This made the bather sweat which got rid of all the dirt on the body. Lastly, the bather plunged into the cold bath and was then massaged with oils and scents.

The baths were a favorite meeting place where people could chat and relax, as well as bathe. Women could use the baths only at certain times, usually in the morning when the men were at work.

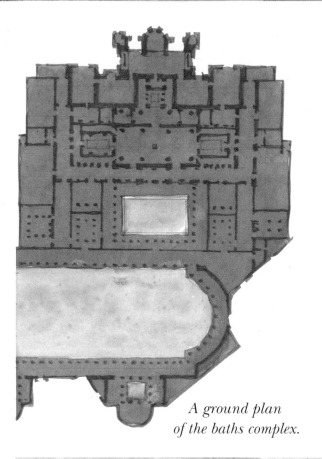

A ground plan of the baths complex.

Hub of the city

Some of the finest buildings from Severus's time were around the forum, which replaced an older forum to the north. It was dominated by the temple which stood at one end and was built on a tall podium or platform 19 feet (5.8 meters) high.

A broad flight of twenty-seven steps led up to the temple. At the top of the steps were pillars of pink Egyptian granite, each one carved with scenes from mythology. They show giants with strange, serpentlike legs battling with the gods, who defeated them with the help of the hero

Janus, the two-faced god.

Herakles (or Hercules). The strange thing is that many of these carvings are signed by Greek stonemasons. We do not know if they worked in Leptis Magna or if the carvings were imported from Greece.

Facing the temple across the forum was the basilica, the large hall where the administration of the city was carried out and courts of law were held. More temples surrounded the paved courtyard of the forum. People would meet in this courtyard to walk and discuss business or government matters.

Lost in the desert

One reason why Leptis was abandoned could have been the desert. The land dried out along the coast and the sands of the desert swept into Leptis and ruined many of the fine buildings. A wall was built to try and protect the city from the sand but it was torn down by invaders as the population of Leptis dwindled. As time went by, the sand completely buried the ruins of the city and the harbor. But though the sand destroyed the city, it also helped to preserve it.

The ruins lay buried until a team of Italian archaeologists, led by Pietro Romanelli, began to excavate it. It was quite easy to remove the sand, and underneath they found many complete buildings, with columns still standing upright and statues in their original positions. More recently the Libyan Department of Antiquities has continued the work.

Among other glories, there is a theater with decorated with pillars and statues, where plays, music, and dancing would have been performed. Wall paintings and mosaics show scenes of life in Leptis Magna, particularly the more gory aspects such as gladiatorial contests.

The ruins are so complete that archaeologists have been able to get a very clear picture of what the city would have looked like at the height of its splendor.

The mysterious arch

One of Severus's most dramatic constructions was the structure known as the Severan Arch. It stood at the junction of the city's two main roads and was really four arches built in a square. This meant that whichever direction you came from, you faced an arch flanked by huge decorative columns.

The arch was built out of limestone covered with marble; this was carved with scenes glorifying the emperor. He is shown with his wife and sons, winning battles, presiding over the courts, and with various gods.

The arch is one of the finest works of Roman sculpture, but the strange thing is, that though many fragments survive, there is no record of why it was built, what it commemorates, or who the sculptors were. Some people think that it was built to celebrate the emperor's return to the city, but that he died before he could return.

A fragment of sculpture from Leptis Magna.

Hagia Sophia

Istanbul, Turkey, c. AD 360–537

*Justinian's magnificent church at Constantinople, with its massive dome,
"suspended as if from a golden chain from heaven," and its gold and silver decorations,
was a symbol of the exotic and fabulously wealthy Byzantine Empire.*

In the fourth century AD the city of Rome, heart of a great empire, was under threat. The empire was out of control, with barbarian invaders fighting the weak emperors and each other for power. However, there was another glorious city, already part of the Roman Empire, which was poised to take the place of Rome.

Byzantium was superbly positioned between Europe and Asia in what is now Turkey. It was surrounded by water on three sides so it was easy to defend. The Roman emperor Constantine (AD 307–337) saw that this old Greek trading town could be turned into a rich and glittering city and become the second capital of the empire.

In six years his architects and builders transformed it into a new city, modeled on Rome, but even bigger and more sumptuous. The elegant buildings and streets were decorated with art treasures from every corner of the empire. The city was completed in 330 and renamed Constantinople, the "City of Constantine."

Rome falls

The ailing Roman Empire was split into two parts in 395, with the western empire being ruled from Rome, and Constantinople becoming the capital of the eastern empire. But Rome was still weakening and it was taken by the Visigoths in 410. Constantinople became the center of what was later to be called the Byzantine Empire.

The Byzantines were the busiest traders of the time. Their merchants came home laden with grain from Egypt, silks from China, spices from the East Indies, gold and ivory from Africa, and furs and wood from Russia.

The Byzantines

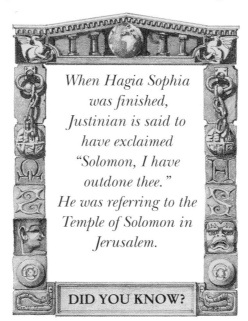

*When Hagia Sophia
was finished,
Justinian is said to
have exclaimed
"Solomon, I have
outdone thee."
He was referring to the
Temple of Solomon in
Jerusalem.*

DID YOU KNOW?

*From windows high in the balconies and dome of Hagia Sophia,
golden sunlight streamed into the church, lighting up the gold, silver, and glass mosaics.
The processions of people would have looked very small in this vast interior.*

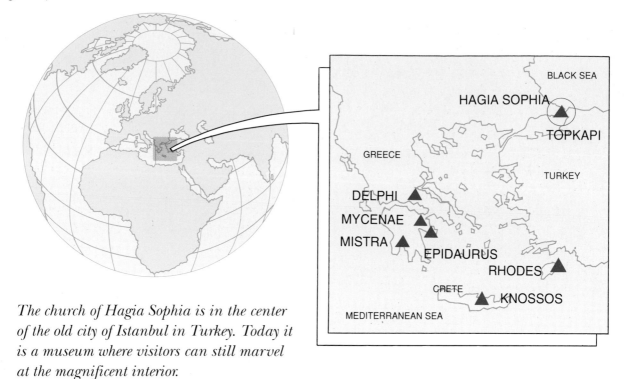

The church of Hagia Sophia is in the center of the old city of Istanbul in Turkey. Today it is a museum where visitors can still marvel at the magnificent interior.

charged a tax on all goods carried in or out of Constantinople. As a result, the city became so fabulously wealthy that its harbor was known as the Golden Horn. Its gold coinage became the international currency used by merchants from as far away as China.

A Christian city

When Constantine became emperor, he took over from a line of rulers who governed with absolute power and were worshiped as gods. Christians were persecuted and killed in Rome, but Constantine realized that Christianity could draw the empire together. In 313 he decreed that Christians should be allowed to worship freely throughout the empire and he later converted to Christianity himself.

The awe-inspiring Christian churches that began to appear were a symbol of the glory of Byzantine art and architecture. The greatest church of all was Hagia Sophia, the Church of Holy

Wisdom, built by Justinian in 537.

Justinian was one of the greatest of the Byzantine emperors. He reconquered the old western empire, and modernized Roman law with his Code of Justinian, which sorted out the complex and often contradictory laws of the old Roman Empire. He also built hundreds of churches. Of these, his crowning glory was Hagia Sophia.

There was already a church, founded by Constantine, on the site of Hagia Sophia. It had been damaged and repaired several times, but it survived until 15 January 532, the day of the famous Nika riots. The ordinary people had gathered at the Hippodrome to cheer on their favorite charioteers at the races. That day the people joined together to rebel against the emperor and his high taxes. The screaming hordes, chanting "Nika!" ("Conquer!"), swept through the city, destroying everything in sight, including the original church of Hagia Sophia.

Justinian and his wife, the Empress Theodora, were strong rulers but this was a test even for them. Justinian was about to flee from the mob but was stopped by a rousing appeal from Theodora: "...for an emperor to become a fugitive," she said, "is a thing not to be endured..." So Justinian and his officials rallied and managed to restore peace to the city.

He continued to rule much as he had before but now he resolved to build the most glorious church the world had ever seen.

Magnificence

More than 10,000 people worked on building the church; Justinian enlisted the help of two mathematicians, Anthemius of Tralles

The monogram of Justinian.

and Isidorus of Miletus, to work out how to support the massive dome. Amazingly, it took only five years to complete the work.

The historian Procopius wondered at the way the massive dome seemed to have no foundation, but looked as though it was suspended from heaven by a golden chain. He was also impressed by the way light flooded into the church through windows in the dome and caught the gold of the mosaics and the silver of the altar screen.

The church became the center of religious life in Constantinople. Most Byzantine emperors after Justinian were crowned there, and it was the scene of elaborate religious services, held with great pomp and ceremony.

Glory to God

A service in any church of this time began with the entrance of the clergy. They would be led by a deacon carrying the book of the gospel. He was followed by the clergyman of the highest rank, the patriarch. Behind him would come the other clergy, singing a psalm.

In the time of Justinian, there were sixty priests at Hagia Sophia, with 100 deacons, forty deaconesses, ninety subdeacons and many other lesser church officials. This large procession walked slowly through the Church to take its place in the sanctuary in front of the high altar.

The congregation would also be large, with men sitting in the nave and women in the galleries. The service would consist of readings from the gospel, a sermon from the patriarch, bringing gifts to the altar, and taking Communion.

Justinian and the Patriarch.

Hagia Sophia

*The builders' main problems
were supporting the dome and
joining it to the square
building below. Both these
problems were tackled together
by building four huge arches
in a square shape to take the
weight of the dome.*

*This arrangement left a
large square floor space inside
the church. The triangular
spaces between the arches and
the base of the dome were filled
with stonework which helped
support its weight. To the east
and west were smaller half
domes. These acted as
buttresses, spreading the
weight over a larger area.*

*Inside the church, the
visitor would first pass
through two entrance halls or
narthexes into the nave. In
front of the sanctuary was the
ambo, a platform where the
gospel was read, which was
big enough to hold the choir.*

*Behind the ambo a raised
walkway led to another open
screen in front of the chancel.
You could see through the
screen to the altar beyond.
This screen was beautifully
decorated with columns and
various details covered in
silver including the initials of
Justinian and Theodora. The
columns were carved with
intricate designs and the tops
of the walls and the roof of the
dome were decorated with
brilliant mosaics made of tiny
bits of glass, often picked out
in gold and silver.*

Power behind the throne

Theodora's passionate words during the Nika riots inspired Justinian to fight back and quell the rebellion, which strengthened his position and made him more powerful than ever before. Justinian realized the part Theodora had played in his success and stated that from now on she was to be joint ruler of the empire. She went on to rule strongly, but who was this dominant woman who influenced an empire?

Theodora was not born into an aristocratic family. In fact, she was of very humble birth. Her father was a bear-keeper at the Hippodrome, the stadium where people flocked to watch the sports of bearbaiting and chariot racing. At the age of twelve Theodora became an actress. In the eastern Roman Empire such women were shunned by polite society.

The historian Procopius tells many stories of Theodora's behavior at this time. She was not a talented actress but she was clever, beautiful, and graceful, and she was seen in the company of many different men, often gentlemen of noble birth. She finally became the lover and later the wife of Justinian, who was then heir to the throne. By now, Theodora's conduct had become very respectable but she was known for her strength and fiery temper.

Justinian with some members of his court.

When Justinian became emperor in 527, Theodora was crowned with him. After that, anyone who gossiped about her past or said anything unflattering about her present conduct was flung into prison and tortured or left to die. Her displeasure was something to be avoided at all costs.

But Theodora also had great courage and determination and many of Justinian's achievements are due to her influence. She did not forget her beginnings and helped the ordinary people, building monasteries, orphanages, and hospitals for them. She also repeatedly pushed Justinian into the actions that made him a great ruler. She died in 548, just before she was fifty. Justinian ruled on for another seventeen years, but he was never to achieve the great deeds that he had with Theodora by his side.

The Empress Theodora.

The fall of Constantinople

The Byzantine Empire survived for more than eleven centuries, but there were always threats from invaders. The Goths, Huns, Persians, Vikings, Arabs, Turks, and Normans all attacked the empire, but it fought off all such invasions until 1071 when the Byzantine army was defeated by the Turks at Manzikert in eastern Anatolia (modern Turkey).

Justinian on a gold medal.

From then on the empire began to decline as more and more territory was conquered. Constantinople was attacked and ransacked by European soldiers from the Fourth Crusade in 1204. The Byzantines under Michael Palaeologus recaptured the city in 1261, but the empire had now shrunk to a tiny area, and it was only a matter of time before it fell to the mightiest invaders of all, the Ottoman Turks.

The Turks were Muslims and they had begun a career of conquest against the Christian world. In 1452 the sultan Mehmet II resolved to capture Constantinople which he called "a monstrous head without a body." He planned his assault for more than a year and in April 1453 100,000 men moved in and besieged the city.

The Turks camped outside the walls, which they bombarded for six weeks. But still they could not break down the walls and enter the city. Then Mehmet decided to stage an attack across the harbor. The city finally fell.

The last emperor, Constantine XI, was killed and Mehmet rode triumphantly on a white horse to Hagia Sophia, which he proclaimed should become a mosque.

Hagia Sophia today

A Muslim shield.

Today, the magnificent church is a museum and visitors are still overwhelmed by its vast interior. But sadly many of its treasures have been lost. The Crusaders ransacked it in 1204, removing decorations and mosaics. When Constantinople fell in 1453 and the church became a mosque, minarets and Islamic inscriptions were added.

Hagia Sophia is still famous for its mosaics, but these are of a later date than the original building. Architectural historians are keen to reconstruct what they can of the original furnishings and decorations, and there is still enough evidence for them to work out what the original designs would have been like.

Mistra

Greece, c. AD 1262–1460

The last, lost city of the Byzantine Empire, perched on a steeply sloping mountain crag. When Mistra was built, the great empire was crumbling all around it, so who built the city and why?

On a craggy hilltop in southern Greece stands the tiny city of Mistra. It is the symbol of the last years of the Byzantine Empire. By the time it was built, the empire had shrunk to what is now Greece and western Turkey. The eastern part of the empire was under constant threat from the Turks, while in the west, lands had been lost to the Normans. And yet this small city would remain the capital of Byzantine Greece for more than 200 years, and become a seat of learning and center for the arts.

Enter the crusaders

In 1204 Constantinople was attacked by the soldiers of the Fourth Crusade and the Byzantine lands were divided up between its leaders. Two of thesé were Frenchmen, from the Champagne district, William de Villehardouin and his nephew Geoffrey. They received the Morea, the area of Greece now known as the Peloponnese. In 1210 Geoffrey became Prince of the Morea.

In 1248, Geoffrey's son William managed to extend his rule into an area of southeastern Greece which was the home of a warlike tribe, the Milengi. They were constantly threatening William's palace near Sparta. What he needed was a new fortress, so he built himself a castle on a hilltop in the nearby Taygetus Mountains. This was to become Mistra.

Legend says that William's castle was named Myzethra, which is the name of a type of cheese, because the hill it was built on was shaped like the cheese.

DID YOU KNOW?

Byzantines fight back

In 1259 the Byzantines, under their emperor Michael VIII, decided to get their land back. They captured the prince and forced him to give up his castle at Mistra. A new Byzantine governor of the Morea was appointed and he arrived at Mistra in 1262. The fortress became the new Byzantine stronghold as they struggled to take more territory from the

The monastery of the Pantanassa is built in the style typical of Mistra, with stone walls and tiled roofs. In Mistra, there seem to be churches everywhere; the best have some of the finest wall paintings in Greece.

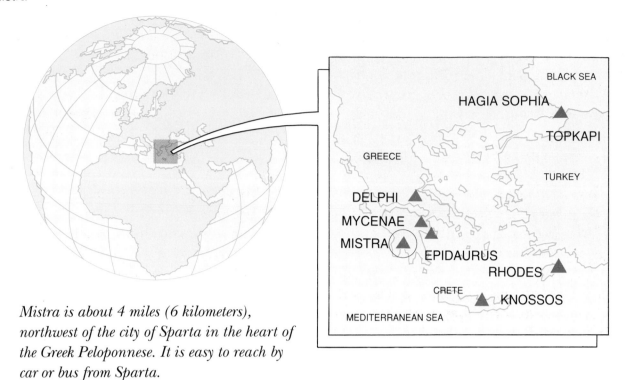

Mistra is about 4 miles (6 kilometers), northwest of the city of Sparta in the heart of the Greek Peloponnese. It is easy to reach by car or bus from Sparta.

Franks. A city with palaces, houses, churches, and monasteries grew up around the original castle.

In 1349, the Byzantine emperor John VI Cantacuzenus made his son Manuel governor of the Morea. Manuel was given the title of Despot. There was very little communication between Constantinople and Mistra, so Manuel and following Despots ruled in their own way, and very successfully.

In 1384, a new imperial family, the Palaeologi, came into power in Constantinople and a member of the new family took over in Mistra. After this, the Despots had more communication with Constantinople, but they were also given more power. The city was enlarged and its artistic and cultural life began.

Hilltop haven

From the Frankish castle which dominated Mistra, walls ran down the hill to enclose the upper city. Within these walls was the Despot's palace, and probably a few nobles' houses. Below this area was another walled section, the middle city, with large churches, nobles' mansions and many smaller houses.

Further down the hill was a third section which was not walled. This was where the poorer citizens lived. The houses and shops in the middle and lower sections were packed together on the hillside, with narrow streets connecting them.

The whole city probably contained 20,000 people. Many of them were Greeks who came to the city for Byzantine protection during the tussles with the Franks.

Scholars and painters

Mistra's period of artistic importance was started by a clergyman called Pachomius who was also an influential official in the city. He was responsible for several of the city's many churches, but his greatest building was the Hodeghetria.

Pachomius wanted to retire to a monastery so he decided to build a new abbey, with a church dedicated to Our Lady Hodeghetria, a Greek title for the Virgin Mary, meaning "she who leads the way."

The Hodeghetria had five domes and a bell tower. Inside there were some of the finest frescoes, sculptures, and marble paneling to decorate any building. Pachomius probably employed craftspeople from Constantinople to carry out this work and persuaded the emperor to award lavish grants to the abbey so that its splendor could be maintained.

Pachomius and other clergymen encouraged scholars, as well as painters, to come to Mistra, and they were supported by some of the Despots.

One of these built the Metropolitan church, or cathedral, which also has outstanding wall paintings.

However, it was the philosopher George Gemistus Plethon who really established the city as an intellectual center (see page 81). He moved to Mistra in about 1407 and was welcomed by the Despot Theodore II.

The end of the Despots

Theodore II had many problems during his long reign, so having welcomed the scholars to his city, he did not have much time to spare for them. The nobility were rebellious and there was the ominous threat from the Ottoman Turks, who by now held much of the original Byzantine Empire. Theodore was heir to the emperor but he died in 1448, so his brother Constantine XI took the throne.

Knowing that the empire could not survive for long, Constantine took the unusual step of being crowned in Mistra rather than in Constantinople. Four years later Constantinople fell to the Ottoman Turks. The Despots lost their power. There were a few years of uneasy relations, while the Ottomans made the Despots pay tribute to them. Then in 1460, the Turks finally conquered Greece. This was the end for Mistra, the last Byzantine city.

The Emperor Manuel II Palaeologus.

Rifts in the Church

As the Turks gained a firmer footing, Emperor Manuel II Palaeologus appealed in 1399 to the western Christian states for help. The two branches of the Christian church, in Rome and Constantinople, had grown steadily apart since the Fourth Crusade in 1204.

Manuel was desperate. He asked for military help in return for a reunion with Rome. His plan did not work, however. The Byzantines would not agree to recognize the hated Rome – even the Ottoman Turks would be better. The West probably did not want the Byzantine Empire to hold power again, so Manuel's plea went in vain.

The Crusades

The Christian world was under serious threat from the Muslim Turks from the eleventh century onwards. This situation inspired two hundred years of Crusades, the expeditions in which Christians from all ranks of society set out to fight for the Holy Lands.

The First Crusade, led by French, German, Norman, and Italian knights, set out in 1096, after the Byzantine emperor Alexus Comnenus had asked the Pope for help to drive the Turks out of Jerusalem. The crusaders managed to capture the city in 1099 after a forty-day siege and immediately celebrated with a bloodthirsty massacre of both Muslims and Jews.

The liberated lands were divided into four states ruled over by crusaders who stayed to defend the territory. Even with their protection, one of the states, Edessa, fell in 1144, which was the reason for the Second Crusade. This, however, was a disaster.

The Third Crusade was called because the Turkish emperor Saladin had managed to recapture every Christian city of the four states, except Tyre. This crusade was led by Philip II of France, Richard I of England, and the Holy Roman Emperor, Frederick I. But the emperor died on the way and the two kings quarreled. Philip turned back and invaded English territories in Normandy while Richard's back was turned. Richard, meanwhile, was left in charge of the crusade and became known as Coeur de Lion, or "Lion Heart," for his bravery. He set up a truce with Saladin which secured territories for the Crusaders and gave them the right to enter Jerusalem.

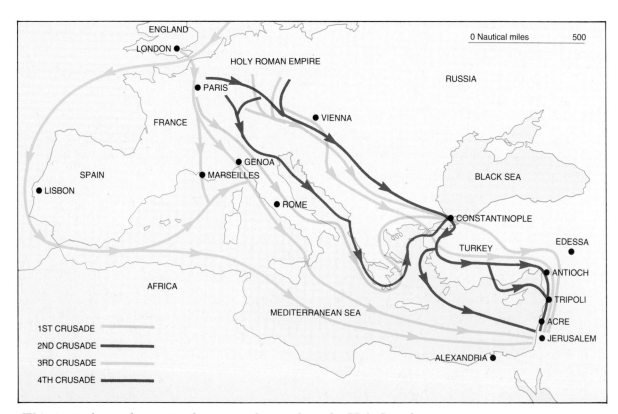

This map shows the routes that crusaders took to the Holy Land.

By now the Crusades were beset by bitter rivalries and greed. This became most evident during the Fourth Crusade when the crusaders looted Constantinople and removed countless treasures from the city. The "knights in shining armor" image of the crusaders had vanished. There were another five Crusades after that but none of them was successful.

A crusader in armor.

Lost in the hills

After the collapse of the Byzantine Empire, Mistra was left to decay. Few people visited the tiny city on the hill. Then, in the late nineteenth century, a French scholar called Gabriel Millet visited Mistra and was fired with enthusiasm for its churches. He wrote a book about the churches, focusing on their wall paintings.

This work inspired other scholars and also the Greek authorities to take an interest. During the first part of the twentieth century, the buildings were restored and the frescoes were carefully cleaned. A more recent history of Mistra by English historian Steven Runciman gives us much of our information about this extraordinary city which survived almost in isolation and yet became a major center of art and learning as well as a focal point of an empire.

Philosopher of Byzantium

George Gemistus Plethon was the most remarkable philosopher in Byzantium. He was a follower of the ancient Greek philosopher, Plato, who wrote a book, *The Republic,* which explained how he thought a perfect state should be run. As Byzantine power declined, Plethon encouraged scholars to look back to Plato's philosophy for their inspiration.

When he opened his academy at Mistra, groups of scholars gathered to discuss many difficult and obscure areas of philosophy and tried to apply them to everyday life. Plethon wanted to see a new Greek state where people were not governed by religion or aristocracy, but were treated as equals in a communal society.

Head of Plato.

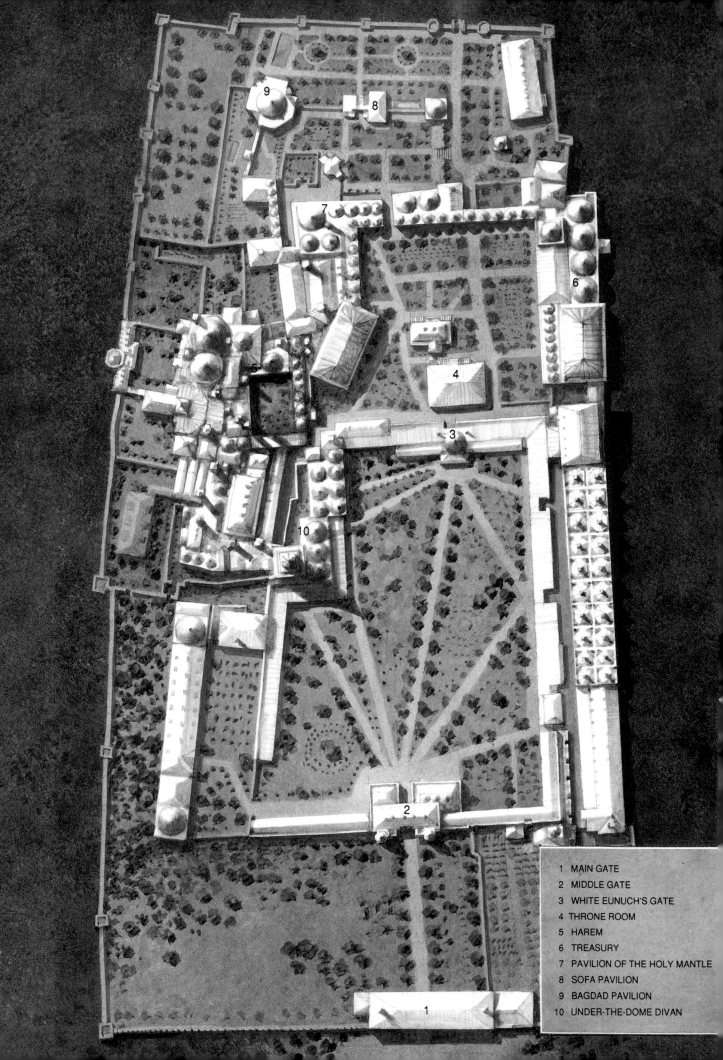

1 MAIN GATE
2 MIDDLE GATE
3 WHITE EUNUCH'S GATE
4 THRONE ROOM
5 HAREM
6 TREASURY
7 PAVILION OF THE HOLY MANTLE
8 SOFA PAVILION
9 BAGDAD PAVILION
10 UNDER-THE-DOME DIVAN

The Topkapi Palace

Istanbul, Turkey, c. AD 1465–1853

*What evil schemes were plotted and murders committed behind
the closed doors of this luxurious palace, home of the fabulously wealthy Ottoman
sultans and their harems of power-hungry women?*

When Mehmet II, sultan of the Ottoman Turks, captured Constantinople in 1453, he found a city that had seen better days. After years of decline, the palaces of the Byzantine emperors were dilapidated and even the great walls of the city were in need of repair.

At the gate to the sea Mehmet built a new wall along the seacoast to give added protection to the city. In this wall there was a water gate flanked by two huge cannons. This was called the Topkapi, or the "cannon gate," so when Mehmet built his new palace nearby, it was known as the Topkapi Palace. It was to become the scene of some of the most evil plots and cruel, murderous acts in history.

Why there?

From the first hill of the city, Topkapi looked out over the great harbor of the Golden Horn, the Bosporus, and the Sea of Marmara so it was in a good position for spotting enemy invaders. It was near the heart of the city, and the great mosque that Justinian's Hagia Sophia (see page 69) had become was nearby, so it is easy to see why this sprawling palace became the sultan's main home.

Topkapi was not Mehmet's only home in the city. He had built a palace on the third hill as soon as he took over Constantinople. But in 1465, when Topkapi was finished, he moved his household there and left the other palace to his dead father's harem.

Sultans who followed Mehmet added to the palace, but it always kept the same character. Most of the buildings were made of stone, though some wooden ones were added at various times. You can still see some of those buildings today. They are low with overhanging roofs and seem to be clinging to the hillside.

One account says that every day the palace cooks prepared 200 sheep, 100 lambs, forty calves, and 200 chickens for the inhabitants of Topkapi.

DID YOU KNOW?

*The palace buildings were separated by large gardens and courtyards.
Most of the buildings were for government officials and palace staff. To the left is the harem
where the sultan and his women had their apartments.*

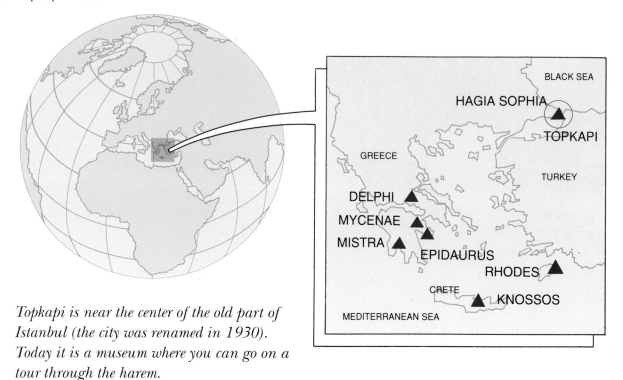

Topkapi is near the center of the old part of Istanbul (the city was renamed in 1930). Today it is a museum where you can go on a tour through the harem.

Outstanding features of the palace were its shady courtyards and lush gardens where the inhabitants of the palace could stroll, sheltered from the sun.

Central power base
As well as being the sultan's home, the palace contained the offices of the government which ruled over the empire. This council of officials was known as the Divan and they met in a small building with domed chambers in one of the palace courtyards. In this building, there were the council chambers, the records office, and the office of the Grand Vizier, the chief official of the sultan. It was the hub of the Ottoman Empire.

By the end of Mehmet's reign, the Ottoman Empire included Asia Minor, Greece, and the Balkans. A century later, the Turks ruled all the lands from the Adriatic Sea to the Persian Gulf. All the decisions about its government were made at Topkapi, often with the sultan watching proceedings from a small, hidden room with a window known as the "sultan's eye."

Visiting the sultan
Topkapi had a magnificent throne room where the sultan received visitors from all over the empire. This was also where he was given news of the empire and heard reports from the Divan.

The sultan also had several reception rooms where he entertained guests. They returned from these visits with tales of the vast wealth of the exotic Ottoman court, with its fabulous jewels and hoards of riches. They spoke of rubies the size of eggs and diamonds bigger than hazelnuts. These claims may be exaggerated, but the sultan certainly was enormously wealthy and loved to show off the fact.

Secrets of the harem
The sultan's women and young children lived in the harem, which was moved to

Topkapi during the reign of Suleiman the Magnificent. The sultan's sons stayed in the harem until they were 11. The eldest son of his first kadin or favorite woman was heir to the throne.

The most important woman in the harem was the Valide Sultan or Queen Mother. Naturally she had great influence with the sultan and was often responsible for planning some of the most cruel schemes. For although life at the palace was luxurious, it was also rife with bloodthirsty plots as everybody tried to gain more power.

For example, there was a tradition that when a sultan came to the throne, all his rivals should be put to death. When Mehmet III became sultan in 1595, all his brothers were executed. The sultan after him, Ahmet I, stopped this gruesome procedure, but instead had his rivals imprisoned in the harem, in an area known as "the cage." Here they were treated as royalty, with luxuries and fine food – everything, in fact, except their freedom.

Exotic decor

The sultan's rooms in the harem were large and well decorated. The biggest was the sultan's hall, a domed room where the sultan and his favorite women would come to listen to music. The sultan sat on a throne, while the women reclined on low sofas on either side of him.

Near the hall were the sultan's baths which were the most magnificent in the palace. Another fine room was a chamber built for Sultan Murat III. The original decorations still survive; the walls are covered with ceramic tiles decorated with abstract designs in blue, red, and green. There are quotations from the Koran, the holy book of Islam, written on some of the tiles.

The chambers would also have had gilt doors, richly colored Turkish carpets, hanging lamps, and fine carvings. We can picture the sultan and his household in their sumptuous robes of velvet and brocade against this background of sophistication and opulence.

Learning to serve

The palace had its own school, set up by Sultan Suleiman the Magnificent, where boys were trained to serve the sultan in the army or as one of his government officials. The boys were slaves captured on Suleiman's many military campaigns and the most talented became the highest officials in the empire. The empire was always governed by slaves specially trained in this way rather than by Turks.

A parade helmet.

When the boys arrived at the school, they went into one of the introductory classes. The talented ones moved to more specialized, advanced classes where they learned to become military commanders, tax collectors, or other government officials. The most talented of all would be chosen for the hall of the privy chamber, where the top officials were trained.

Life in Topkapi Palace

Here, Sultan Mehmet II is seen with some of the inhabitants of the palace outside the main gate. They include harem women, a mullah or Islamic scholar, soldiers, the Grand Vizier (on horseback), palace officials, entertainers, and a visiting emissary. The palace was always swarming with life, as richly robed officials went about their business and servants bustled around, carrying out their bidding.

Topkapi Palace was guarded by the sultan's troops, the janissaries, who had rooms in the main gate and kept watch on everyone entering or leaving the palace. They were very powerful and made sure they remained so by starting rebellions if they felt their power was diminishing.

The palace was a sprawling collection of buildings which were added to as necessary. Life

was given some kind of order by arranging the buildings in groups around courtyards (see page 82). The first courtyard contained storerooms, the arsenal for weapons, workshops, the mint, the bakery, and the hospital. The servants also lived here. The Divan's chambers were in the second courtyard. The palace kitchens ran along one side of this courtyard, so there was a continuous bustle of servants going to and fro.

Each kitchen had a special function. One prepared food for the Divan, another for the harem, and a third for the sultan. In the center of this courtyard was a garden of fruit trees – almonds, figs, lemons, oranges, and pomegranates. The throne room and the palace school were in the third courtyard. The harem was on the left-hand side of the second and third courtyards.

Behind the veil

The women of the harem were the sultan's slaves, but they could become very powerful. In fact, one particular lady was probably responsible for having the harem moved to Topkapi. She was Roxelana, the second kadin of Suleiman the Magnificent.

After the death of the Sultan Valide, Roxelana plotted and schemed to gain power for herself. Her final triumph was to become Suleiman's wife. To have the harem moved to his main palace would make her power even greater.

A woman of the harem.

A woman who was a favorite of the sultan and gave him a male child could become a kadin, with her own apartments and servants. For the rest of the women life must have been very boring. They were not allowed to go out of the harem and saw no one except the sultan and the eunuch guards. The most they could hope for was to achieve some position of status, such as Keeper of the Jewels or Mistress of the Wardrobe, or to become a confidante of a kadin or the Sultan Valide. The atmosphere in the harem must have been a

Suleiman the Magnificent

The Ottoman Empire reached the height of its power during the reign of Suleiman the Magnificent, who came to the throne in 1520. He was a cruel but brilliant leader and he was helped by his friend, Ibrahim, who became his Grand Vizier.

The Turks were constantly at war with Christian Europe and Suleiman, with Ibrahim by his side, continued to expand his empire. His first victories were to capture Belgrade in Serbia and the island of Rhodes (see page 55) soon after he came to the throne.

Then he turned his attention to Vienna, the capital of the Hapsburg Empire, in Austria. This was the last city barring the way into Christian Europe. However, Suleiman failed to take the city because his guns were

bogged down in the mud by heavy rains and he could not use them in his three-week attack. Yet Suleiman and Ibrahim managed to persuade everyone that they had won a victory.

As his successful campaigns increased, Suleiman became hugely rich. He dined from silver plates encrusted with jewels and wore sumptuous clothes. He enjoyed his luxurious lifestyle and left the government of his empire to Ibrahim.

As Ibrahim became more powerful, he also became very vain; he boasted that he had more power than the sultan himself. This made him many enemies and one of these was Roxelana. She convinced Suleiman that Ibrahim was stealing his power and betraying him. One evening, Ibrahim went to dine with

constant buzz of plotting and scheming as women vied to catch the sultan's eye and thus become more important themselves.

Ibrahim, Suleiman's ill-fated Grand Vizier.

Topkapi abandoned

By the time Suleiman died, the Ottomans had built up an empire of thirty million people. Until then, the sultans had been cruel but strong leaders. Now everything changed. The sultans who followed Suleiman were weak men who were only interested in enjoying themselves.

The empire managed to survive until 1922 but it never again achieved its earlier greatness and its power began a serious decline in the late seventeenth century. Topkapi remained the main palace until 1853 when Sultan Abdul Meçit I moved to a new palace at Dolmabahçe on the Bosporus.

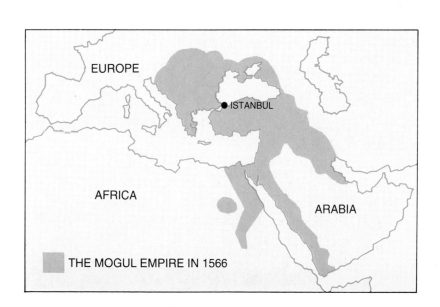

EUROPE

● ISTANBUL

AFRICA

ARABIA

THE MOGUL EMPIRE IN 1566

Sultan Suleiman the Magnificent.

the sultan: the next morning his body was found at the palace gate – he had been strangled.

Now Roxelana plotted to have the sultan's eldest son, Mustafa, murdered so that her own son, Selim, could become the sultan's heir. In 1553 the sultan had Mustafa strangled as he watched from behind a curtain. When Suleiman eventually died in battle at the age of seventy-two, Selim became sultan, but by then Roxelana was dead.

Where to go

Although pictures will tell you a lot, it's much better to go to a museum and look at all the things that archaeologists have found from a vanished civilization. You will get an even better idea of how a people lived and worked and what they thought was important by looking at the statues, jewelry, pottery, and other remains.

Some museums have special visiting days when they let you actually touch these ancient things and examine them. Often school visits are allowed special access to items not usually on display, if they are studying a particular period or culture. But **always** check the opening days and times before you try to visit a museum to avoid disappointment.

Museums

The Art Institute of Chicago, Michigan Ave. and Adams St., Chicago, IL 60603, (312) 443-3600.

The Brooklyn Museum, 200 Eastern Pkwy., Brooklyn, NY 11238, (718) 638-5000

The Cleveland Museum of Art, 11150 East Blvd., Cleveland, OH 44106, (216) 421-7340

Dallas Museum of Art, 1717 N. Harwood, Dallas, TX 75201, (214) 922-1200

The Denver Art Museum, 100 W. 14th Ave. Pkwy., Denver, CO 80204, (303) 640-2295

Indianapolis Museum of Art, 1200 W. 38th St., Indianapolis, IN 46208, (317) 923-1331

Los Angeles County Museum of Art, 5905 Wilshire Blvd., Los Angeles, CA 90036, (213) 857-6111

Metropolitan Museum of Art, Fifth Ave. at 82nd St., New York, NY 10028, (212) 879-5500

The Minneapolis Institute of Arts, 2400 Third Ave. S., Minneapolis, MN 55404, (612) 870-3000

Museum of Fine Arts, 465 Huntington Ave., Boston, MA 02115, (617) 267-9300

North Carolina Museum of Art, 2110 Blue Ridge Rd., Raleigh, NC 27607-6494, (919) 833-1935

Philadelphia Museum of Art, 26th St. & Benjamin Franklin Pkwy., Philadelphia, PA 19130, (215) 763-8100

Phoenix Art Museum, 1625 N. Central Ave., Phoenix, AZ 85004-1685, (602) 257-1880

Portland Art Museum, 1219 S.W. Park Ave., Portland, OR 97205, (503) 226-2811

Seattle Art Museum, 100 University St., Seattle, WA 98101, (206) 625-8900

Virginia Museum of Fine Arts, 2800 Grove Ave., Richmond, VA 23221-2466, (804) 367-0844

Specific sites

Tarxien – most of the finds from the various sites on Malta and Gozo are on display in the National Museum in Valletta, which is the capital of Malta.

Knossos – most of the finds from the various digs at Knossos and the other palaces of Crete are in the museum at Iraklion, the capital of Crete.

Mycenae – most of the finds from Mycenae are in the National Archaeological Museum in Athens.

Delphi – there is an excellent museum just beside the entrance to Delphi which includes the world-famous statue of the Charioteer.

Epidaurus – there is a museum at Epidaurus which contains many finds from this site. Other items are in the National Archaeological Museum in Athens.

Rhodes – nothing remains of the Colossus; all we have are ancient descriptions of it. The Archaeological Museum in Rhodes Town displays items from the period when the Colossus was built.

Leptis Magna – most of the best finds are in the museum in Tripoli, the capital of Libya.

Hagia Sophia – there are no finds from Hagia Sophia. The best collection of relics of the Byzantine Empire is in the Archaeological Museum in Salonika, Greece.

Mistra – most of the frescoes are still at Mistra.

The Topkapi Palace – the whole complex is now a museum with guided tours through selected areas. There is also a display of the treasures of the Sultans, including an emerald as big as a softball.

Further Reading

Archibald, Zofia. *Discovering the World of the Ancient Greeks.* New York: Facts on File, 1991.

Burrell, Roy E. C. *The Romans.* New York: Oxford University Press, 1991.

Coltrell, Leonard. *The Bull of Minos.* New York: Facts on File, 1984.

Euripides. *Plays One/Euripides.* Intro. by J. Michael Walton. New York: Methuen, 1988.

Evslin, Bernard. *Heroes, Gods, and Monsters of the Greek Myths.* New York: Four Winds Press, 1967.

Homer. *The Iliad of Homer.* Trans. and intro. by Richmond Lattimore. Chicago: University of Chicago Press, 1951.

James, Simon. *Ancient Rome.* New York: Knopf, 1990.

Jenkins, Ian. *Greek and Roman Life.* Cambridge: Harvard University Press, 1986.

Lewis, Raphaela. *Everyday Life in Ottoman Turkey.* New York: Putnam, 1971.

Palmer, Leonard Robert. *Mycenaeans and Minoans.* Westport, CT: Greenwood Press, 1980.

Perowne, Stewart. *Roman Mythology.* New rev. ed. New York: Bedrick Books, 1983.

Powell, Anton. *The Greek World.* New York: Warwick Press, 1987.

Rice, Tamara Talbot. *Everyday Life in Byzantium.* New York: Putnam, 1967.

Willetts, R. F. *The Civilization of Ancient Crete.* Berkeley: University of California Press, 1977.

Index